Absolute Relativity

The

Theory of Everything

Edward Johnson

Edward Johnson

Published in United States of America by Edward William Johnson 2012

Copyright © 2012 Edward William Johnson

Edward William Johnson:
ISBN-10: 1470157950
ISBN- 13: 978-1470157951

DEDICATION

Mother Violet Beatrice Johnson, Father William Henry Johnson Lt Royal Navy
Andrew Mitchell Lt/Cdr Royal Navy, Ashley Thom Lt Royal Navy. Sisters Jennifer
Charlotte & Rita Winifred.

CONTENTS

For
Fabiola
&
Thomas Edward Johnson

The Speed of Light
is not absolute
it just seems that way, it is
the tail of the lion

Euclid 323–283 BC

The Elements

Sir Isaac Newton 25 December 1642 – 20 March 1727

Philosophiæ Naturalis Principia Mathematica

Albert Einstein
14 March 1879 – 18 April 1955

Special & General Theory of Relativity

Hermann Minkowski
June 22, 1864 – January 12, 1909

Minkowski spacetime

Al-Jayyani 989-1079

The Book of Unknown Arc's

Edward Johnson

ABSOLUTE RELATIVITY
THE THEORY OF EVERYTHING

The expression for all the dimensions of 'Spacetime'
including *space-time*:

$$(Ut, x, y, z,) + t$$

Where 'Ut' is the local rate of expansion
of the universe.

Where Hubble visible *space-time* is:
$$x, y, z, + t$$
't' is temporal time

The expression for the Energy/Mass ratio:
From Einstein's $e=mc^2$ becomes $E=m(ut)^2$
Where the quantum of energy released is determined by the
value of local 'Ut' expansion.

The instantaneous destruction of the entire universe
will be determined and triggered:

$E=m(ut)^2$ when the value of Ut falls below the Atomic
Zero ground state to cause the Standard Model dissociation
causing 100% conversion of all atomic mass to energy.

The value of (Ut) above will be in the range: >1' < 299,000km/s.

The God Particle and the Higgs Field does not exist, it is nothing
more than the expansion of the universe with an absolute symmetrical
time framework of '0', with a local velocity of 300,000km/second

1. **Absolute Spacetime** of the entire universe has a time symmetry of '0' or '1' which has been preserved from the moment of origin.
2. **Has one** dimension.
3. **It has a unit** expressed as (ut) meaning Absolute Spacetime. It extends the Euclidian spacial units (ut, x, y, z,).
4. **The fabric** is a 1 dimensional field construction, which may be built from a field of energy strings.
5. **This dimension** is constantly expanding in all directions, meaning co-flow, counter flow including all transverse angles.
6. **The speed of light** is determined by the rate of expansion of this dimension.
7. **The expansion** of this dimension is creating New Space constantly around us since the Big Bang or origin.
8. **The continuous** creation of New Space enables and is essential for all things to move anywhere in universe including all quantum and sub quantum & biological matter.
9. **It is impossible** for anything to move faster than its ability to create New Space - determined by its rate of expansion.
10. **Atomic clock** measurements confirm the presence of Absolute Spacetime 'Ut'. **(a)**. they measure the ***Proper time as Zero*** when co-moving with it at a speed of 300,000Kms. **(b).** they detect Absolute Spacetime ***Expansion Velocity*** as being 300,000kms when co moving with it when ***Proper time = Zero.***
11. **The same** measurements will be detected by atomic clocks anywhere in the local Space. Different measurements will be detected in other parts of the universe which will detect a variable expansion of New spacetime.
12. **That the entire** universe will disappear in less than 1 second of proper time if the entire universe stops expanding or slows.
13. **That all** the events that have happened, is happening or will happen here on Earth or any other place or planet occur at the same time with reference to Absolute Spacetime of '0'.
14. **All cosmological events** past, present and future also all occur at the same time frame with reference to Absolute Spacetime.
15. **The rate** of expansion is not uniform across the entire universe.
16. **Magnitude of mass:** it determines the meaning of invariant and relativistic mass. & determines black hole morphology.

THEORY OF EVERYTHING

1. **This** dimension provides the singularly most important dimension that it provides the ability and framework for all the other spacial dimensions to exist, including the passage of temporal proper clock time.
2. **Einstein's** Theories of Special and General Relativity are determined by the limits and conditions of this dimension.
3. **Should** the rate of universe expansion slow, will cause the reduction in the speed of light and will cause the immediate extinction of all suns in the entire universe.
4. **Should** the rate of universe expansion increase, will cause all suns to irradiate substantially more energy and the immediate destruction of all biology in the universe.
5. **It determines** and determined all physical knowledge and information ever caused in the universe since its beginnings.
6. **It is causing** the continuous generation of cosmological matter behind the periphery of expansion.
7. **It enables** anything and everything to have motion in the universe. And enables exchange of information between species.
8. **The universe** is expanding around us and through us minute for minute throughout our entire lives.
9. **It determines** the speed of radio transmission and the absolute achievable & theoretical velocities of any item and particles.
10. **It determines** the amount of energy released in an atomic explosion, energy released from suns and the ability of energy release for power and digestion of food in our bodies.
11. **It provides** the ability for all biological matter to move, a space for flowers to open their petals, and enables butterflies to visit them, and the wind to move a leaf.

DETECTING ABSOLUTE SPACETIME THE 4TH DIMENSION

How can you detect something which you can't see? Well there are 100's of scientific examples where we have done this. A good example is Galileo. Everybody on the planet who has ever lived by the sea is aware of tides, the regular Ebb and Flow of the sea with tireless (historically magic) regularity. We have only really understood this phenomenon for a few centuries thanks to the intuition and observation of Galileo in modern astronomy. He determined that it was the rotation of the moon around the planet which was causing this effect. The result of this amazing insight caused him to be put under house arrest for the rest of his life, as it went against the official views of the Roman Catholic beliefs and condemned him for heresy! Sometimes when something is obvious to the individual the institutions will reject it – this is the way of humanity. Institutions very often have blinkered eyes and do not want to realise the truth as it upsets their status quo understanding of things, and not prepared to accept a change. It might be just too inconvenient!

Then the late Sir Isaac Newton extended this understanding by suggesting it was the effect of the moons gravity which was pulling the water towards it. All that makes sense to us now, but at the time, and not really so long ago this knowledge was not available to us. Prior to Galileo nobody had any idea what caused it.

Then Einstein who then extended the above ideas of Newton, stating that this gravitational effect was causing a virtual curvature of space-time between the Moon and the Earth. This concludes our basic understanding of what causes the Ebb and Flow. We have what is known as a hysteresis effect. Meaning the action follows the cause. In the above case the sea is being pulled towards the moon. That is the Cause, and the Action is the tide.

In order to determine and prove the existence of the notion, the 4th spacial Absolute Spacetime dimension we have to look for evidence in a similar way. Ok so how? If we think about Atomic clocks, nobody really understands the time dilation as presented to us by Einstein, not even he understood this phenomenon, discovering the action without knowing its cause. In this book I propose it is the cause of the 4th dimension and I think a powerful explanation.

What we have to do is look at nature and evidence of things happening which don't seem to have any explicable reason, or at least an understood one. The tidal regularity is a typical example. So here is another example for the scientists to muse over. Sun spots! The sun is very nearly 1,000,000 miles in diameter. That is quite a big picture to hang on the wall! We can use that massive profile to detect cosmic irregularities. And think of it as a scientific detector, in the same way as Galileo was considering the frequency of tides then found a relationship linked to the Moon.

We observe what is known as Sun spots – a phenomenon which we are all aware of. If you research this you will discover that observations of sun spot date back to 364B.C. thanks to the Chinese astronomer called Gan De. We also know that Sun spots have a pattern of 11 years. Nobody has the faintest idea why – synonymous to the observed tidal effects centuries ago. Also with today's modern instruments we can detect believe it or not historic Sun spot activities recorded for us in tree rings. We now have a record of Sun spot frequency of an amazing 11,400 years. That in itself is quite a staggering bit of scientific investigation.

Einstein would be the first to agree with me that if the speed of light has a diminished velocity, one of the first things to happen is that it would affect the amount of energy liberated in an atomic bomb, nuclear fusion or fission – thanks to his $E=mc^2$ equation. So what I am proposing is that we consider that the Sun is a huge scientific detector of movement and variation of Spacetime passing through it. Or simply to try and detect that it exists. Henceforth then determine the speed of light is it a reliable as we think it is? In any moving field (universe expansion in this case), it may be subject to variances no matter how small. We will be completely oblivious to any such tiny variation, particularly if we are not looking for it. .

If the velocity of light, or as I propose the expansion of the universe does have local variations. It will affect Einstein's equation above. If the variations are to cause a downward change in velocity it would cause our Sun to be dimmer as the available energy liberated will be lower! Again, Einstein would agree with me on this point. Therefore, what I am suggesting is that the Sun spots as observed may well be due to tiny variations of the expansion of the universe causing the light speed to drop and hence small areas on the sun to be dimmer! This is an astonishing idea, and I can already here the science community around the world having a belly laugh at such a notion. But on the basis that the community has no idea what is causing them – someone has to present something. Even if they put me under house arrest for the rest of my life for heresy! It is currently believed the spots are nothing more than internal magnetic storms – they may be right? – But why?

This is just an idea and I think if we do accept the concept and applicability of this 4th dimension of Spacetime, we have to discuss it and find ways of proving that it does exist or does not. And develop new experimental techniques to detect it, and most of all look at nature and the cosmos in a different way as the clues exist waiting for us to find them. From my point of view the performance of atomic clocks and *proper* time dilation is sufficient evidence for us to take it seriously.

Black Holes

Has anyone proposed that black holes may be black holes NOT just because they are super dense – but they are rotating at near half light speed or more and behaving like an inside out HADRON? Later in the book you will please note that I propose that a mass (its variant mass) of a body is determined by the universe expansion moving through it. It the case of a '**black hole**' its finite mass is reacting with it. **Causing** its '*variant mass to super densify*' (*gets heavier*), because it is rotating so fast in it! Then transmit into space all that momentum energy, and not a simple thermal decomposition. It could be a physical interaction with spacetime 'ut'. The inferred **mass and gravity,** causing it become a mass to energy convertor. And its dense armature - (the dense ball at its centre in this case) rotates through the movement of spacetime expanding at 300,000km/s to facilitate this.

What I propose essentially is that the dynamics involved with black holes is the physical interaction rotating in the constant expansion of the universe 'ut'. And this interaction causes its mass to increase substantially. The black hole cannot gain this extra mass! The momentum so high with respect to its rotation and velocity of universe expanding through it. Then exchange the attempt at variant mass gain by shedding it off as transmitted energy into space.

To put it another way it wants to get heavier, heavier and heavier because it is spinning so fast – in an environment which is also moving fast through it and it can't.

It is under constant pressure because of the resident dynamics: the relationship of velocity of rotation, the resident mass, and the speed of Ut passing through it to get heavier. And it cannot!

Then causes the huge *virtual* space-time curvature according to Einstein's GTR / Euclid x, y, z dimensions. And hence a huge effect upon all materials in the immediate and distance from it. It is not the fact that it is heavy in isolation that causes entire galaxies to rotate around it. It is the physical relationship of its starting density, its rotation and more important the capacity of the expansion of the universe moving through it. It may never have sufficient attractive gravitational power in its own right to attract millions of stars 100,000 light years away from it. The reason we think in these terms is because of Albert Einstein GTR. His basic theory is completely correct. But a tiny object the size of a cosmological pin head would never be able to have sufficient density act as a sink for entire galaxies. It is the relationship it has with the spacetime 'ut' which is causing this affect.

It is creating an artificial *enlarged gravity effect in space-time* existing in a plane of visible space around it (x, y, x), and determined by the velocity of spacetime..

The spark plug analogy diagram below provides a simple insight & explanation.

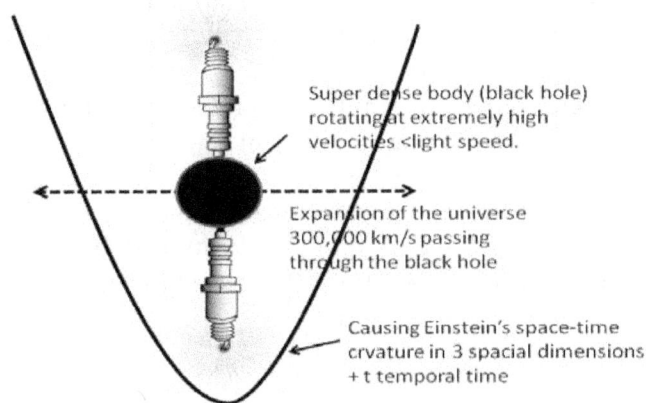

Super dense body (black hole) rotating at extremely high velocities <light speed.

Expansion of the universe 300,000 km/s passing through the black hole

Causing Einstein's space-time crvature in 3 spacial dimensions + t temporal time

Please excuse my use of such simple illustrations. I think it gives and immediate idea of the concept which is being presented here.

Because, the super dense body is rotating so fast and the 4th spacial dimension 'Ut' is passing through it. The body is reacting – then obliged for its relativistic mass to densify (*increase its weight*). It struggles to do this so it emits substantial energies which we observe.

Its options are: **1.** Get heavier or **2.** Get rid of energy which it should it use to make more mass. **3.** It can't gain mass, because if it does it will have to reduce its diameter. **4.** It can't reduce its diameter because it will have to spin faster. **5.** It can't spin faster because of the interaction with spacetime passing through it, causing it to convert the extra mass into energy!

It remains in a state of equilibrium. And can be thought of as a continuous mass to energy inverter. The cause of its huge virtual gravity is a relationship of its dimension, rotational speed and the velocity of spacetime passing through it which limits and determines its achievable gravity/mass & energy characteristics.

The morphology of a black hole is to achieve a constant equilibrium and is self limiting in proper time at the rate **1.** The speed the velocity Ut is passing through it. **3.** its diameter **4.** its starting mass. **5.** its rotational velocity. **6.** The attraction of additional material which is being fed into it such as neighboring stars etc – which then have to be expunged as energy, having reached its equilibrium with Ut. It is impossible to get any heavier. Should the velocity of Ut change either upwards or downwards then under those new conditions will change its state of equilibrium. Then modify the production of gravity by an increase or reduction and with it the subsequent transmitted energy.

It will continue to attempt further densification so long as the body, **A**. continues to have such a huge rotational velocity and **B**. the rate of expansion Ut remains the same i.e. 300,000km/s. **A:B** are causing the resultant gravity.

Sun spots!

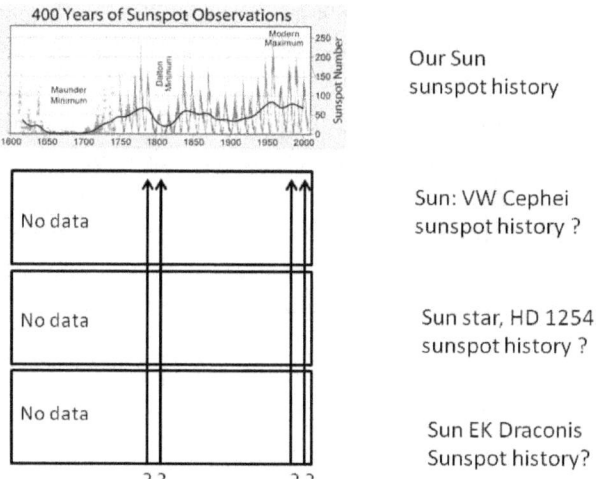

Our Sun
sunspot history

Sun: VW Cephei
sunspot history ?

Sun star, HD 12545
sunspot history ?

Sun EK Draconis
Sunspot history?

Sketch showing the frequency of sun spots of our Sun.
The empty boxes below for 3 other suns which we have
been observing have no information.

The sketch above shows the recorded frequency of our sunspots. We observe other suns, 3 of which named also above. We know they have sunspots, and some cases so substantial they cover 11% of that suns area. So it is an obvious question how do those sunspot frequencies compare with ours? I have looked into this briefly but have not been able to find any data. If data does not exist then maybe NASA should put this on their work list. If there is an information link between sunspot activities then that would be an amazing discovery!

Then the science community will have opportunity for new investigations. From the point of view would support the presence of spacetime 'ut' and have me dancing in the street with happiness. As obviously something is affecting all suns simultaneously and happening between them!

Bearing in mind the proposed observations and measurements would have to be corrected for light speed between the various suns we are monitoring. From my point of view it may be indicative of spacetime acting between them. I will wait to see if the science community will investigate this! In any event I think I have illustrated that perhaps we should find ways of determining its existence and applicability in modern science. If it does exist it will open new books of science and confirm this Theory of Everything.

4th single spacial dimension propagating information through Spacetime

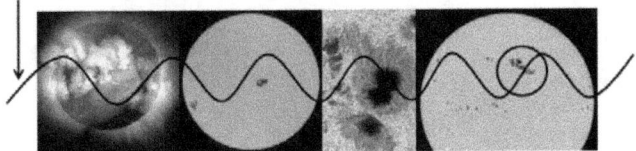

We should compare the occurrence
of sunspot activities of observable suns . To
determine if they happen at the same time.

The image above illustrates some random sunspot activity of our Sun. If we study other observable Suns we would know if they all happened simultaneously. Should they indeed happen at the same or similar moment, this would indicate that they are being affected by something acting between them. In this case Absolute Spacetime Ut, expanding between them.

1 INTRODUCTION

Spacetime! Don't panic it is only a word, and of course should have a meaning. The insight presented in this book will endeavor to explain it – if only to prevent me from forgetting what I think it is. Hopefully, you will be able to share my amazement of it. When we hear the word spacetime or see this word we immediately look for something else to entertain our minds, or launch into thoughts about Einstein's theories which, in the main leave's most of us quickly exhausted - just in thinking of it. Please don't recoil. His work and that of Sir Isaac Newton are mentioned in these few pages, and of course some references – only for purposes of making a helpful explanation of how we currently understand things in the world and universe around us - and a means to comprehend how the universe has an absolute spacetime which up to now has never been properly explained to us.

Spacetime may be currently misunderstood by the scientific community and the rest of us who are influenced by them. In this book you will be able to appreciate its existence and the meaning in the universe, in a completely new way. And be introduced to a fresh way of considering physics.

Sir Isaac Newton used the term '**Absolute Spacetime**' centuries ago. In coining this phrase he did so, not because he understood what it was but had a belief that something must exist in places where matter does not. Matter meaning, an atom, pebble on the beach, a planet, sun or galaxy etc. The description, the form and or location of such matter does not matter! He once said *"What, is there in space where matter does not exist"?* He is known as one of the world's greatest scientists and his work further developed by Einstein centuries later. Einstein once said *"I can see further only because I can stand on the shoulders of giants"*. He was referring principally to the work of Newton. However, this phrase was in fact Newton's phrase, and he was referring to the work of even earlier philosophers such as Euclid for example who wrote his laws of spacial coordinates, (*he split up the space around us into 3 dimensions of* **length, width and height** *expressed* as **x, y, z** *'The Elements'* in 300 B.C. That it quite amazing in itself, where a high tech device at that time would have been no more than milking stool! He is known as the father of geometry and his book has been print in all that time. And is today 2nd in the list in terms of popularity after the bible! I call it a book but of course in 300 B.C. his theories would have been written on papyrus of course. His geometric interpretation of the world still holds good today, and Newton and Einstein used his 3 dimensional coordinates in their own work much later.

Spacetime is a word which means space and time which is simply incomprehensible to most of us. **'The standard definition'** more or less states from a technical dictionary!

Spacetime (or space-time, space time, space-time continuum) is any mathematical model that combines space and time into a single continuum spacetime is usually interpreted with space as being three dimensional and time playing the role of a fourth that is of a different sort from the spatial

The biggest problem with understanding what spacetime is - is that the very idea of it is confused. The people who define it can't really define it in a way which can understand it.

We have given a *'something'*, a word without really understanding what that *something* is! And regularly confuse it with Euclid's 3 dimensions and the concept of proper time. Sir Isaac Newton could not understand it or Einstein. Whereas, Newton believed that spacetime exists as some form of incomprehensible empty space which should have some kind effect on matter. Einstein completely discounted this idea, as it had no applicability to any of his Theories – namely his Special Theory of Relativity and later his General Theory of Relativity. However, even though he discounted the ability of spacetime as separate entity, he knew there must be an explanation for it, but evaded him for his entire life.

Before, we go any further - as we are using words which will cause confusion. It is necessary to explain and separate the meanings of **'Spacetime'** and **'space - time'** right away.

1. **Spacetime** is an easy expression, and according the author has a simple definition, being a single dimension, which occupies the entire universe including us. (To be further defined later in this book with some revelations). It forms the PRIMARY DIMENSION of the universe and is expanding all around us and though us. It must not be confused with the description below. It does not apply the use of proper clock time. (or **'proper time'** being your wrist watch). This is nothing more than a man made unit of time – not natures. We use it as a means of pure convenience to move through x, y, z spacial dimensions to get to the shop before it closes!

2. **Space and time, space-time, & space time continuum etc** are expressions for 3 spacial dimensions and the use **proper** clock time in the observable universe. That is everything in the visible part of the universe where we live.

3. **Proper time** or proper clock time is a frame of reference of any event which is measured in units seconds, minutes and hours, months etc. And the units are of course fixed to the rotation of the Earth for convenience. They are used in the 3 spacial dimensions and cause confusion when referring to spacetime.

I hope that helps. Please keep them separate in your mind for the purposes of reading the following explanations. This book will introduce a means to combine them later which defines the concept of 'The Absolute Relativity' and Theory of Everything. String Theories and or M Theory are not really consideration in this book, although it does not discount the idea that Absolute Relativity the 4th dimension 'Ut' explained here may well be built from energy strings of some form or other. It is not really important for this explanation and will only cause confusion. Should more dimensions exist or be found outside what is being presented - is hereafter thought of as being lesser importance. On the basis that String Theorists may find that every atom in the universe is a dimension in its own right and worthy of yet another dimension to be added to their equations. Having said that there is an opportunity to include string theory constructs into spacetime being proposed, which scientists may wish to argue about that later.

2 BACKGROUND BASICS

Before we move on to some explanations regarding Absolute Relativity let's consider a little information relating to Einstein's Special Theory of Relativity to ensure we can separate the world of Einstein from the effects of spacetime in our thinking. Einstein completely discounted the idea of spacetime and was totally excluded in any of his theories.

Einstein's Special Theory of Relativity has a home and is located in the 300 B.C Euclid's world of his 3 spacial dimensions + proper time = x, y, z + t. Don't worry about all these characters, they simply express **height**, **width** and **length**, which all of us are aware of and take for granted. They are also illustrated by the following space diagram:

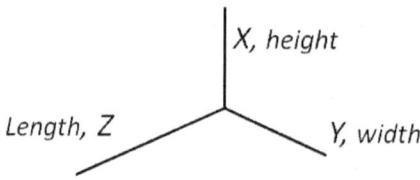

There is nothing really complicated about that, and is our home, and home for Einstein's theories! 3 simple dimensions in space defined as three coordinate vectors x, y, z. Which are used to define volumes by giving these coordinates numbers i.e. 1 inch, 1 mile, 1 light-year etc a dimensional block of our <u>space</u> located in the Big <u>Space</u>. It is essential to understand this simple framework. These 3 spacial dimensions form and confine our entire existence of the world and Space as we know it, or see it.

Having easily understood Euclid's x, y, z, dimensional definition then we have to say Ah! What about time? As obviously we must use clock time or also known as proper time in moving from X to Y, or Y to Z etc, to measure time as we move in these framework distances until we are quite bored by it. We move in an x, y, z space framework and proper time passes! Daily we are most aware of this as we simply cannot be late for work and we know that moving in the world of $x, y,$ z we need a clock to achieve a satisfactory outcome which hopefully is not being sacked because we are not late. We all take this additional (temporal proper time) for granted, and we can easily determine in our mind how much time we need to carry out a particular task or achieve movement in our knowledge of living in $x, y, z.$ $+ t$ framework of life.

In the x, y, z world where we live we add an extra **virtual** dimension of clock time. I say *virtual* time, as it uses arbitrary time units when measuring movement through $x, y, z.$ Our units of time just happen to be linked to the rotational speed of the Earth — we all know that! So when we move through $x, y,$ *and* z we are measuring our movement as units or part units of Earth rotation. I.e. seconds, hours, days etc.

In the true sense clock time is not a dimension; we use it just for its convenience and synchronized to the Earth rotation — seeing we live here. We can now define our existence expressed as 3 dimensions, *(x, y,* and $z.$ And within these simple 3 spacial dimensions gives us a definition of space and the ability of proper time we are able to record our movement as we move through them. We are quite happy to live out our lives from birth to death within our everyday experience of these planes *and* not need any more dimensions to improve our lives.

In 1907 a gentleman by the name of Herman Minkowski very early on, recognized the importance of Einstein's Special Theory of Relativity and stated Time is a separate temporal dimension - hence our 3 daily dimensions x, y, z $+ t.$ This is called Minkowski space or Minowski's spacetime. Then since that moment the concept of **space - time** and **Spacetime** have been confused together, and continue to confuse. Minkowski's spacetime philosophy is based upon the application of the 3 known vectors of Euclidean space as applied in Einstein's work. That being the real visible space, confined by 'The Box' dimensions $x, y, z,$ $+ t$ which we all live in. Neither of them have any claim on the applicability of physics to actual Spacetime' which will be described in the following chapters. However, on the basis that Euclid invented the concept of space in 3 spacial dimensions $x, y, z,$ then Minkowski has technically contributed to Euclid's earlier work and extending its relevance in the world in which we live. We now realize that it is obvious and maybe if Euclid had a wrist watch 300B.C. maybe he would have given us an expression for 4 dimensions not 3 (the 4th being known as temporal time). However, I am sure he had a means to determine temporal time but he never included this in his x, y, z spacial law. In his case it might have been the sound of a cockerel in the morning, the sun at midday, his wife calling him for dinner or perhaps a sundial these would have been his temporal proper time references.

3
THE FORGOTTEN DIMENSION

Einstein's relativity theories as stated earlier occur in the 3 easily understood Euclidian spacial dimensions. His predictions relating to the curvature of space and time etc have to be confined by them. However, some may say it is a violation of Euclidian laws of 3 spacial dimensions!

Why? Simply because Euclid made no allowance in his 3D straight line space world, to permit the inclusion of a curve! Einstein determined in his General Theory regarding the curvature of Space. - is using curves in a 3D field which according to Euclid must only exist as straight lines!

So what! ? Well he is technically violating the 'Euclid rules' as it only made a provision of 3 dimensions – all straight lines and no rule for any curvatures! How can you apply a curve into a framework which only permits straight lines? This is a dimension conflict.

Einstein would never consider such nonsense, on the basis that the box volume as defined by the physical lengths of the sides provides a subsequent space = volume. And we can do whatever we want within that volume – which means inclusion of curves at will. And we all do it, but we are taking something for granted – how can we do it if there are no rules in Euclidean space to permit the inclusion of curves? For example and again Euclid's space is x, y, z not $x, y, z + Curve$!

You will doubtless be dubious and I am being pedantic. Let me explain further. In the period 989-1079 A.D, a gentleman Muslim being an Arabic mathematician called Al-Jayyani had a problem. He alike his fellow religious believers needed to make certain predictions in order to fix the time for religious prayers and festivals etc. These predictions where based upon cosmological occurrences such the location of the moon reference to the Earth. So he determined what is now known as Spherical Geometry in his treatise '*The book of unknown arcs of a sphere*'. This is what is known as Non Euclidian space, and technically extends the thinking of Euclid yet again – so let's now consider the extended Euclid law *(x, y, z, t, Sg)* Where **Sg** is Al Jayyani's ***spherical geometry*** added to it. So now mathematically we can now include curves and circular activity into the straight line Euclid universe box of straight sides. To advance my earlier comment in order to live out our lives we now use the **x, y, z**, box and (**t**) for temporal time, and a law for spherical geometry (**sg**). At last physics has a set of easy rules and defines all the necessary coordinates for us to exist in. Whether we are aware of these rules is not very important to us, as we carry out our daily lives. But everything we do must involve them.

Concept of <u>Spacetime</u>

We are still missing something. When we define a volume which may be a cube, a sphere or any other geometric shape, it is <u>not</u> 'Just' the definition of 'Empty Space or the object' – it is also the definition of a volume of **Spacetime** which exists in its own right. Which, we are not aware of. However, it will still exist, and up to now never thought seriously as a **separate dimension**. Henceforth we need to regard the measurements and Space as yet another set of rules which further extends the Euclid list earlier stated.

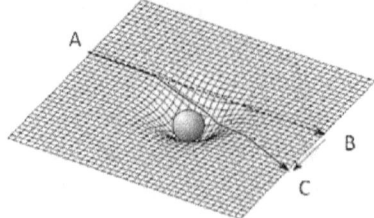

The Einstein field (*3D box frame*) on the previous page is located within a standard Euclidean 3 dimensional linear framework of space + Spherical Geometry as stated above. It does not show any consideration for Spacetime which, Einstein completely discounted as he had no way of knowing its applicability in the above framework. In simplicity the line A – B is a line with a nil curvature in absence of any mass in its vector field line. The line A-C experiences a virtual local space curvature due the presence of a mass appearing at the bottom of the hollow. In reality the Euclidian space- time will not actually appear like this, it is a simple mathematical realization to assist our understanding of the gravitational effects around the mass and the curvatures are subject to the numerical value of its mass – i.e. how heavy or dense it is.

Actually, this consideration is not really important as we have a far more interesting place to visit and think about. But it is necessary to establish in our minds what is happening or predicted to happen in Euclidian space determined by Einstein's General Theory of Relativity. His theories only relating to the Relativity & Deformation of 'space – time' in the volume world of (x, y, z +t). It provides **no allowance** for the presence of **Spacetime**. Einstein referred to it as curvature of **space-time**, which indeed it is but not **Spacetime.** These 2 terms have been confused since 1905. At best the concept of Spacetime completely ignored by science – as having 'No Applicability' by courtesy of Einstein's influence on us. As genius as Einstein was and remains incredibly important to us, he completely negated the most important dimension of all as without it none of the other dimensions can exist!.

In considering the above Einstein space-time curvature of Euclidian space in this way is an introduction of the most important dimension which has never been really considered since Newton. It adds a further Euclidian unit: (**Ut,** *x, y, z, t,*)**.** To confirm our comprehension of this new idea please note the illustration on the next page:

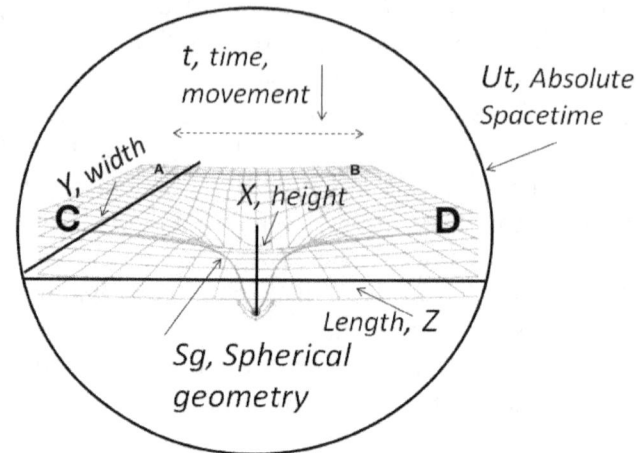

X, Y , Z, t, Sg ,Ut *This represents all the applicable dimensions and extends the Euclidian list to enable and explain our ability to exist in the world and universe*

$$(x, y, z, + (ut) = 4 \text{ real spacial Dimensions}$$

Plus: Non Euclidian units of dimension = '*t*' man made proper clock time and '**sg**' Al Jayyani and Pythagoras for curves and circles.

All we have done is to include a further dimension representing **Absolute Spacetime 'ut'** to the Euclidian list. You will understand the function and applicability of the others. And this new '**forgotten primary ut**' dimension is now represented by a great (*circle*) a sphere encircling the other secondary inner dimensions. No further dimensions will be added and the above diagram is all we need to understand the more interesting ideas to follow. In the diagram on page 9 it shows a simple affect of a single mass or planet causing the virtual curvature of space & time (the simple x, y, z + t dimensions). All we have done is add to it an extra dimension (*the 4ᵗʰ Real Spacial dimension and actually = the* **Primary dimension 'ut'** *)*. Up to now it has been completely ignored as having a nil value or contribution to the laws of our physical world. Hereafter, it will be explained why this is a scientific shortcoming, and indeed is the missing dimension and the answer to everything.

Shopping list of all <u>Spacial</u> Euclidian dimensions:

1. X = height
2. Y = width
3. Z = length

Shopping list of non Euclidian dimensions:

4. t = Temporal proper clock time.
5. sg = Spherical geometry necessary to include curves inside square boxes.
6. **ut = Absolute Spacetime** = the new 'primary' 4th spacial dimension being discussed in this book. Without it none of the above shopping list can exist! That is how important it is.

The only ones we are concerned about for the purposes to explain Absolute Spacetime and the Theory of Everything is 1, 2, 3, **and mainly 6.**

4 TIME SYMMETRY IN THE UNIVERSE

At the time of origin the BB, we are lead to believe it occurred in a tiny Space – a microscopic dot. And may have given us more than just expanding void of Space and *baryonic* matter *(mass and particles),* and home for our subsequent universe. At the time of its origin if indeed it did happen by means of a tiny spot, it happened within 1 contained miniscule instant of time.

Therefore one may presuppose the diameter of that supposed microscopic spot shall have the same time then, as it does now – with its huge diameter. The time of that 'Event' is locked and fixed within its spherical framework. Furthermore, that time shall exist across its entire framework as an absolute reference point of fixed <u>Spacetime</u>.

Keeping things simple - let's think of a simple analogy. A chicken lays an egg, and that egg has a shell. Let's further imagine that this egg is the size of an atom. Let's also imagine that this special egg and shell is elastic in all directions. The egg shell expands and its contents expand with it. The contents may become somewhat rarified *(thin),* but that is not an important consideration. The eggshell and contents will continue to exist as an egg – it is only bigger. The really important thing to consider it that the moment of origin when the egg was laid remains exactly the same no matter how huge the egg becomes. As illustrated in the following diagram.

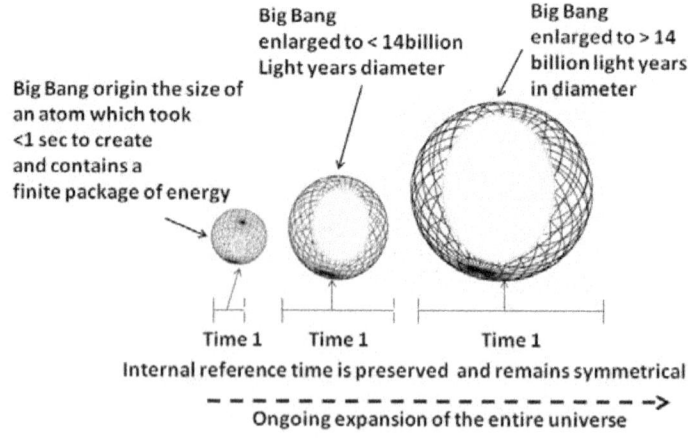

Big Bang origin the size of an atom which took <1 sec to create and contains a finite package of energy

Big Bang enlarged to < 14billion Light years diameter

Big Bang enlarged to > 14 billion light years in diameter

Time 1   Time 1   Time 1

Internal reference time is preserved and remains symmetrical

– – – – – – – – – – – – – – – – – – –>
Ongoing expansion of the entire universe

Please carefully consider the illustrations to assist with the reasoning. Some scientific opinions advise that the expansion of the universe is similar to an expanding balloon. That is essentially correct, however, it is doing more than just expanding. It is preserving the very time moment when it was created. That is the preservation of the original event time within itself: Then this event is stretched over an ever widening volume. In the illustration below the event time remains symmetrical, meaning: it has not changed, and is the same time and constant from its original atom sized package to now something of the order 14 billion light years+ in diameter.

However, the time in this case is not proper clock time – it is spacetime, it is important not to think of it as a clock which has simply stopped. We are obliged to use the word time as there is no other word to communicate the action which, is the preservation of the moment of genesis. Let's look at another analogy on the next page.

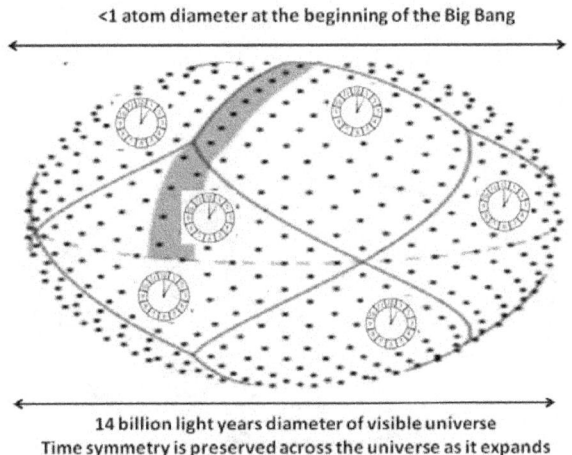

<1 atom diameter at the beginning of the Big Bang

14 billion light years diameter of visible universe
Time symmetry is preserved across the universe as it expands

The above is a way of looking and understanding this effect. The picture represented by the dots is the universe as it expands. The proper clocks illustrated (*a red herring in this case*) indicates the same moment irrespective to their location. It also illustrates exactly the same moment of time whether the universe is 1 atom in diameter and the present day some 14 billion light years wide. Proper time is meaningless in this case, it is not measuring the universe internal framework maturing – the clock images inserted are only used to illustrate that everything within the containment of the entire universe, no matter what its size, exists and shares the same event moment. All that is changing is the increasing size and volume of the universe – **not** the passing clock time of the original moment it was created. Scientists have tried to illustrate this but unfortunately unable to, mainly on the basis that their investigation is always centered on quantum particle behavior. As mentioned in the introduction to this book that the dimension I refer to as **Spacetime 'ut'**, is totally separate from 'space – time' **and has <u>no</u> measurable <u>units</u> as it is a <u>dimension like a width, height or length</u>**. And as soon as one uses the quantum world *(atoms and subatomic particles as references)* to help explain anything in spacetime it is impossible on the basis that spacetime is a totally separate entity and has no quantum material in it. And as soon as one uses these materials to make sense of anything one immediately returns to the world of Einstein's General and Special Relativity. The universe is nothing more than an expanding field of an absolute moment, which is moving in all directions at once. That is across its diameter and its periphery – its entire fabric is moving and expanding. And of course has been moving since the very second it was created approximately 14 billion years ago.

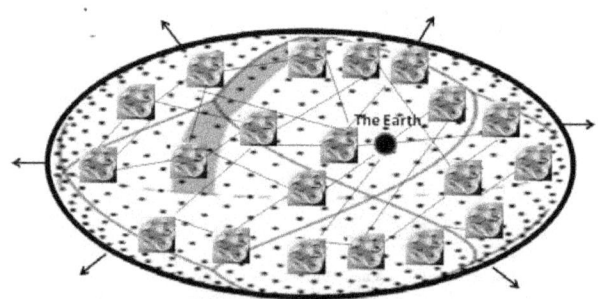

The universe comprising a lattice field
of superstrings. Synonymous to the ultra thin expanded chicken egg.

However, there may be a case for Superstring materials which may be responsible for the construction of the fabric of spacetime? If so this would be roughly similar to our analogy of the expanding chicken egg scenario referenced earlier. I am sure you could imagine how rarified that egg density would become. Thin it maybe but still exists as an egg, and in moving through it you would not be aware that it was an egg at all!

So if we are prepared to consider that the superstring lattice of the universe – constructs **our Spacetime** we must not forget it is not a stationary fabric. It is **extremely** mobile and in our visible universe
(called the Hubble Zone) is moving at the speed of light 300,000kms/sec. And not only that, is moving in *All* directions up, down, sideways, in and out all at the same time simultaneously, just as it did when it was expanding from the size of an atom. And you may in fact consider that the Big Bang has not been concluded it is still happening! In the same way you may consider an atom bomb. If you could survive standing at its centre when it went off! You would say ouch that's an atom bomb! A 2nd person who is 10 miles away would also say ouch that was an atom bomb a small time afterwards. A 3rd person and 4th person at further distances away would also account for the same event. However, your account of the atom bomb going off would be the actual detonation event – whereas the 2nd and subsequent witness's, if they do not witness the actual detonation would say, as far as they were concerned it detonated when it struck them! And likewise all other witness's until a diameter of 14billion light years distant. It is still expanding with unbelievable ferocity. As far as the force is concerned it is just as powerful at the time origin as it is now. And everyone standing anyway in its path would say ouch. The bomb is still exploding.

One can also visualize this expansion as being driven from within its internal fabric as a form of pressure – or by an immensely energetic periphery pulling & displacing everything within its fabric outwards.

As it expands in all directions simultaneously it is creating **new space** in our region of the universe at the speed of light. More specifically – '**New space**' means '**New Spacetime 'ut'**, don't forget it is a dimension like width, length or height. Is being created in the very room or place where you are reading this book. But as the egg is so thin you are completely unaware of it. You are sitting in a room which is located inside an expanding egg made up of a potential superstring lattice-field. Expanding around you at the speed of light, but also too towards you – from above and below you – every direction – and being created second by second at the speed of light. The universe is getting bigger, in the distance from your eyes to the hands which are holding this book! All around you, day and night - since the day of your birth till your visit to the grave. It has been doing this for 14 billion years!

And we are completely oblivious to it, we think of it up to now that is: – where is the edge? of the universe and how quickly it is moving away from us as if it is disconnected from us! But the expansion is everywhere not just at the invisible periphery. We live and move in its expansion! We witness the expansion everyday but completely aware of it. But it influences everything around us – especially light.

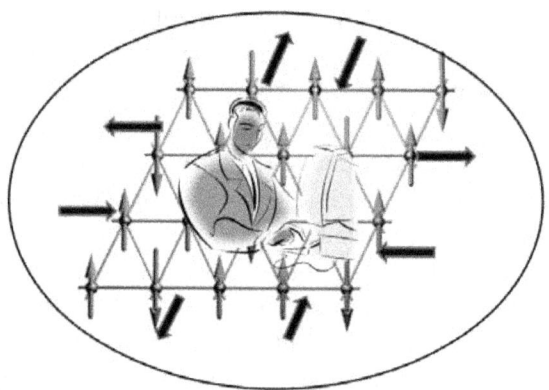

Spacetime 'ut' –with its dimensional lattice expanding
and closing in all directions.

The above diagram illustrates the 4th dimension 'ut' as a lattice field (the ever expanding egg not just at the periphery but across its entire volume). Were the elements of it are moving in all directions. Up, down, contra flow, co flow and all radial angles simultaneously. New space is being created at the accidental velocity of 300,000 km/s.

5 GENERAL THEORY VS ABSOLUTE SPACETIME

The illustration below shows a typical Einstein's General Relativity field showing curvatures in regular Euclidian space-time caused by the presence of mass and gravity. It has nil effect upon the dimension Absolute Spacetime '**ut**' framework.

Einstein's laws of mass/gravity/space-time curvature of General Relativity are obliged to inhabit this separate dimension of spacetime **ut** despite his denial of its existence.

The effects of mass & gravity do not cause any or virtual deformation to Absolute Spacetime. In fact the reverse is true, by means of descriptions later in the book. It is actually defining the behavior of physical laws which exist within in it. Not the other way around. The physical constants which Einstein and Newton use are determined by the conditions pertaining **directly** to the **rate of expansion** '**ut**' of the universe in our locality. That is causing the observed and measured behavior of **everything.**

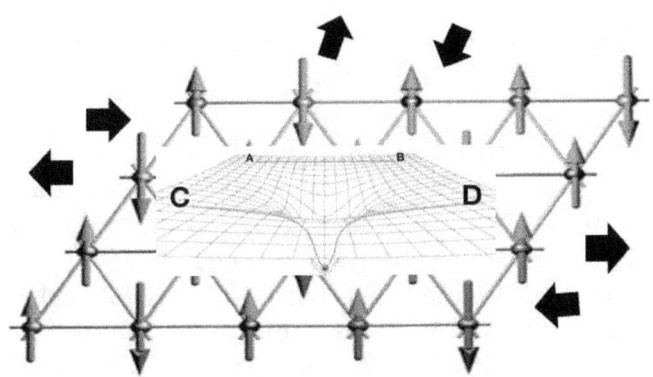

A General Relativity mass causing the curvature of
'*space – time*', but having a no effect upon the
dimension of '*Absolute Spacetime* lattice field.

6 ABSOLUTE SPACETIME UT VS THE SPEED OF LIGHT

This particular chapter is going to present some interesting notions, and may cause you to reconsider everything differently in the world around us. The following notion is a completely new way of looking at how physics and the universe work.

Man has been experimenting with light and inventing different experiments to determine the speed of light for centuries. Recently due to advances with measurement devices we know more or less exactly its finite speed, other characteristics and behavioral patterns such as refraction, reflection, gravitational lensing, the duality that it exists as a wave and a particle etc. Despite our extensive knowledge relating to the science of light and its paradoxes etc, we do not know the reason **what determines its velocity**! The science of man simply states that it exits as a constant and that is that! And in my opinion this is unsatisfactory. For instance we know what determines the speed and propagation of sound energy such as air and solids etc. But in the case of light - no explanation is provided why travels as fast as it does. We know we can slow it down – so why can we speed it up? There is a missing explanation.

We all take the speed of light with its physics reference 'C' totally for granted. – It happens to be 300,000kms, without any further consideration as to why! Well, here is the possible answer, to a question which science have never properly pursued. Then having understood this new concept - that the speed of light is actually being determined by some other force – and not the ability of a photon to determine its own velocity. Most if not all of Einstein's important work is centered upon the concept that it is universally constant. However, I have to add that this is completely satisfactory for the purposes of everyday comprehension of the visible physical dynamics around us. But should he have found out what determines the speed of light then perhaps he could have given us the Theory of Everything?

Light we know it is an extremely small particle - if indeed it is a particle however, whether it is or not is of no importance in this consideration. Albeit, the nature of light which may consist of tiny mass less photons are actually existing and interacting in **4 spacial dimensions** not just the **3!** This is an amazing revelation and brings something completely new to our world of understanding nature and physics! Let me explain more, thank God light is visible and it enables us to see the world around us, and indeed distant cosmological entities – suns, galaxies planets etc. But is it doing more than making things visible, it is a means to illustrate visually its ability to move in 4 dimensions simultaneously? The following diagrams illustrate this idea.

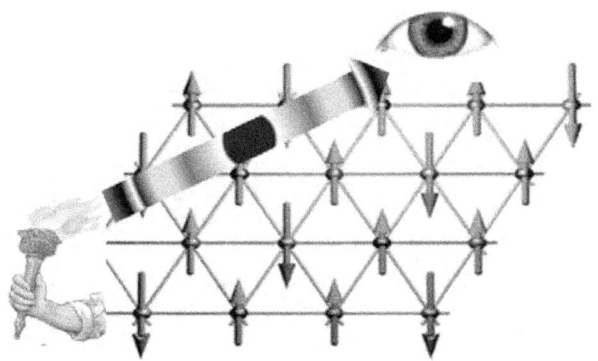

The speed of light is not determined by its own physical ability – it is predetermined by the expansion of the dimension of Spacetime.

Ignoring what is happening in the other 3 spacial dimensions it is obvious to us, as we witness it every day. It can be appreciated that as the universe is expanding it is creating invisible **new space** around us, all the time, second for second constantly. The ability of light to move in this dimension is completely determined by the ability of the universe to create **new space**. We have no experience not in the 14billion years of its existence of things moving in a static universe! It just appears static around us as we have no way of detecting its enlargement at the incredible rate of 300,000km/sec. Other than by looking at nature and especially light. In 1 second of proper space-time units the amount of **new space** created is 300,000 kilometers! That is a lot of new universe space in such a short period of proper time! For instance in the time it has taken me to swallow a sip of tea the universe has enlarged around me to that distance! Hence the speed of light, moreover **if the expansion** of the universe was 800,000kms – then of course the speed of light would be entirely the **same velocity**. There has never been given any cause or reason for the speed of light – simply because no one has ever presented a supportable idea. Newton and later Victorian scientists used a phrase called the 'Aether', - thinking that it was a medium which caused the transmission of light in vacuum. Their thinking may have been on the right lines. However, the concept being presented here is completely different and new - and is based upon the expansion of **new universe space** everywhere in the visible universe. Another sketch to covey this idea:

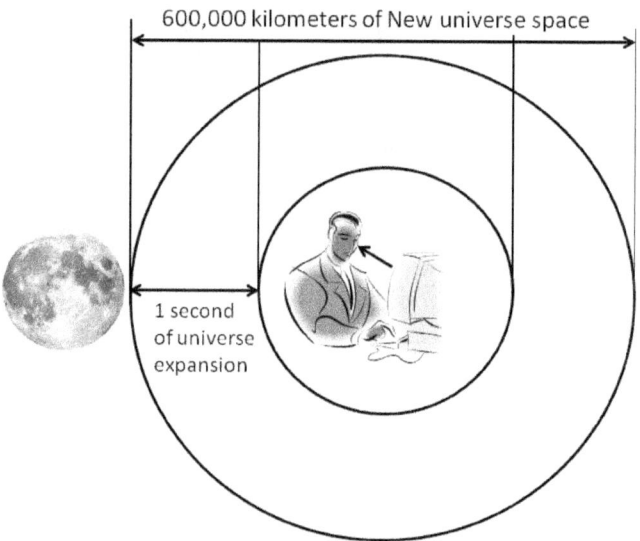

As the computer operator is typing away above - new Absolute Spacetime 'ut' is being created around him. As shown in 1 second of proper clock time light from his monitor would have reached the moon. He is completely unaware of this occurrence and such a thought does not enter his mind. It also does not enter his mind that the light being transmitted from the computer monitor is able to reach his eyes because of this expansion between him and it! If the universe **was not expanding** in this 4th spacial dimension, then the ability of any light to move whatsoever would be **completely impossible**. As illustrated by the following sketch.

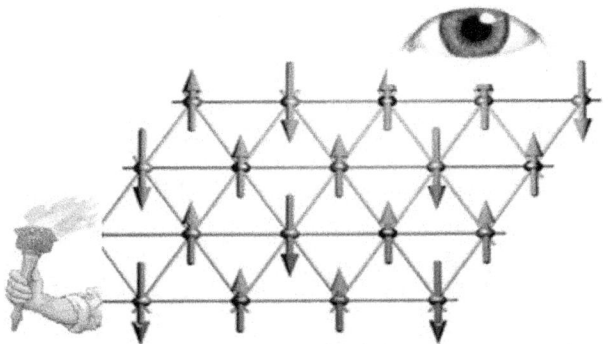

If the expansion of the universe is Zero, terminating the creation
of New Space, then light cannot reach the eye of the observer,
and its velocity will be '0' Km/sec

In this scenario the eyes of the observer would then be immersed in a world and universe which is completely opaque. Which means not even black!, as there would be absolutely no observation of any light or any objects immediately around us, mid distance and at vast distances nothing would be visible! This alarming outcome is even more alarming as it would also affect every physical law which is based upon the speed of light (actually *'the expansion of the universe'!*).

For instance in the case of Einstein's famous equation **E=mc²**. The amount of energy releasable from a mass of an object or particle in this equation is the weight of that mass multiplied by the speed of light, multiplied by the speed of light. Should the universe (4th dimension of Spacetime 'ut') not be expanding and not creating **'new space'** as described'. Then the amount of energy released from 1 gram of mass would equal 0. Let me illustrate this incredible notion by using the sum below:

The universe expanding at 300,000km/s:

1. $E = mc^2 = (300,000 \text{ km/s})^2 \times (\text{Speed of light squared})$.
$= 89,875,517,873,681,764 \text{ J/kg} (\approx 9.0 \times 10^{16} \text{ joules per kilogram})$

So the energy equivalent of **one gram** of mass is equivalent to:
25.0 million Killowat/hours ($\approx$25 Giga Watt Hrs)

The entire power consumption of France is typically 101,700Mwatt/hr

If the universe is NOT expanding:

2. $E = mc^2 = (0 \text{ m/s})^2 \times (\text{Speed of light squared})$.
$= 1 \text{gram of mass} \times 0 \times 0$
$= 0$

Therefore the amount of energy available from 1 gram of mass is
Zero!

No. 1, above is based upon the speed of light which is consistent in our region of the universe = all the visible space. So we are we fortunate that the universe is expanding (creating new Spacetime) with this particular velocity of 300,000 kilometers per second locally.

No. 2, Above the situation is based upon (no new Spacetime production) hence a velocity of 'Zero'. This would have an extremely alarming effect upon everything! Not just because 1 gram of mass will not release a single Watt (unit *of power*) of energy. All life as we know it will cease to exist immediately. The world and the entire universe around us will be a very different place indeed – and science fiction writers may feast on this idea for a long time to come!

Furthermore, the ability of **all the Suns** to react to produce heat and light will immediately stop. The ability of light and any energy to be transmitted will stop. Any ability to even move a tea cup will be impossible! In essence **if** this 4th spacial dimension '**ut**' explained here **does not expand** then the entire envelope of the universe will simply freeze solid! And will simply provide an appearance in a video where the objects in it are completely motionless. However, such a video would be completely impossible as light would not be able to move for the cameraman to see the objects. And! Electron clouds around atoms would also be rendered motionless! – And! if electrons can't move around atoms – we have no atoms! Hence the entire universe would disappear in less than one second!!

Then people will respond to this argument and say: "What about the motion of the planets"? "gravity"? "Surely they will maintain their Newtonian laws of Motion?, and Einstein's laws of gravitation and space – time curving? But again the answer will be no! Nothing, absolutely nothing will be able to move. Quite simply all movement is a license provided to all matter – huge planets, sub atomic particles and especially light. The 4th dimension must create new Space moment for moment and is essential for anything to move in the other 3 spacial dimensions which we live in.

7 ATOMIC ARMEGEDDON
END OF EVERYTHING

Atomic cohesion is caused by its ability for it to vibrate. The atoms comprise sub atomic particles known and defined by the Standard Model. These are the building blocks which comprise all atoms.

If the ability of movement by virtue of the expansion of the universe is cancelled or slowed these particles will not be able to remain as a cohesive entity in the Euclidian x, y, z, t dimensions. They will immediately lose all there atomic cohesion necessary to construct a single atom anywhere in the universe. Every single atom will be pulled down below its so called '***zero-point field***' and cease to be an atom anymore.

This would occur within the entire universe in less than one second! **This is how the universe will destruct itself**. When you split an atom it causes the release of substantial energy. Every atom in the entire universe will be forced to release its energy simultaneously! On page 22 you can see how much energy is available from 1 gram of matter (based upon the expansion speed of the universe). The approximate total mass in the visible universe is **3.35×10^{54}** Kilograms. If you have the will you can work out how much energy that would liberate! You can use $E=mc^2$ – but I suggest using a lower number than 300,000ks on the basis that the atomic cohesion will be lost at some arbitrary figure <C speed of light. Should you wish to calculate it I would suggest using ¾ known light-speed just for fun? On the guess and assumption that at this speed atoms will lose their glue ability and explode. It would be a memorable explosion! In fact it would be a **NEW 'BIG BANG'**! Then create an envelope of **New Spacetime** with its own New Absolute Relativity. A new expansion and brand new Primary Dimension of Spacetime necessary for Euclid's x, y, z, $+ t!$ **to exist in it.** All over again!

The diagram below shows the Standard Model List of sub quantum particles which construct atoms. If as stated the value of 'ut' reduces to such a level it will modify the ability of their interactions. In doing so will cancel their ability to behave as atoms and will break apart liberating their energy as they revert back into a state of pure energy. All mater will simply vanish including us!

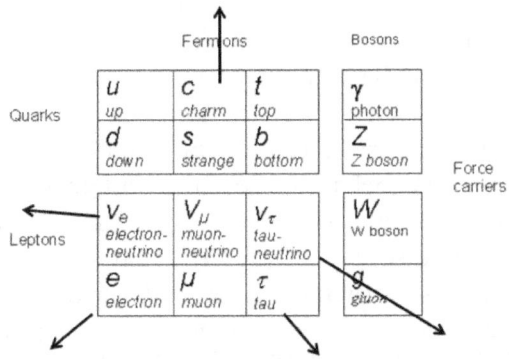

Atomic particulates of the Standard Model which are the construction components of atoms will lose their cohesive interactions. All the atoms will disintegrate simultaneously and will be converted back into pure energy.

If the Spacetime 'ut' expansion slows and disables the ability of atomic cohesion, not a single atom will exist in the entire universe. Revert back to a cloud of energy and end history of the current universe in the three dimensions + temporal time.

THE 'NEW BIG BANG' AND THE DISTRUCTION OF EVERY SINGLE ATOM IN THE UNIVERSE

Higgs Field

It is proposed that the Higgs Field which the science community seeks is '**not a field**' as presented by Peter Higgs it is the expansion of the universe and its velocity. If we look at Max Planks famous equation below it can be seen the speed of light is essential in the equation. Both notable scientists accept that the speed of light is a constant. For the purposes of everyday science yes it can be considered as such. However, in order to appreciate and understand the ability of light, atomic and sub quantum behavior we must understand that the speed of light is not a constant.   On the simple understanding that the conditions of the universe expansion may change. The universe is a dynamic system and is moving. The rate of this movement (the expansion) may be subject to change – then at which point if and when that happens will **instantly affect the speed of light.** In doing so will change all the values of 'C' in every scientific equation and affect all atomic behavior instantly. This is what the scientific community does not understand.

$$f = \frac{m_0 c^2}{h\sqrt{1 - \dfrac{v^2}{c^2}}}$$

Replace 'C' with
Ut^2 universe
Expansion velocity

de Broglie Frequency m_0 = rest mass   h = Planck's Constant

$$m = \frac{\hbar}{Rc}$$

Replace 'C' with
Ut universe
Expansion velocity

m = mass, R = Radius C = speed of light

Planck's constant above (symbol h). A constant that relates the energy of a photon or atom to its frequency. It has the value 6.626 069 × 10^{-34} Js. It is named after the German physicist Max Karl Ernst Ludwig Planck (1858–1947).

This equation is used to determine the so called Zero Point Field which determines the lowest energy state of an atom. It also shows the energy state of a photon above under the current conditions of speed of light. If however you insert a theoretical value for 'C' above and replace it with a different value representing the expansion of the universe then ALL conditions change. This essentially is my argument for ATOMICIDE the extinction of all atoms. The mechanism is the expansion of the universe, it slows to a point where the Zero Point Field is so low all atoms **will lose their ability** to be atoms. Their cohesive interaction will simply not be possible. **This is the Mechanism for the Creation and Disassembly of atoms.** It is not a Higgs Field!

$E=mc^2$

We are all familiar with this equation ingeniously presented to the world thanks to Einstein, however its evolution does have a long history going back centuries and including a scientist called **Lavoisier.** He was guillotined during the French Revolution on the basis that he worked as a tax collector for the King of France. The Revolutionists turned down an execution pardon on the basis that the revolution does not need a genius! He originated the symbol 'C' for the speed of light. In latin it has a meaning Celeritas meaning "swiftness."). He was finally pardoned 1 year later – a pity they took lost his life!

The symbol Celeritas in this case is a fixed velocity constant of light and of course used in Einstein's famous equation. But if you can appreciate that the speed of light is determined by the movement of the universe similar to Planck's constant does not give us the full picture. It should also be changed to the equation $E=m(ut)^2$. **The "swiftness" being the expansion rate ut of the universe around us – not light!**

Then we have a new equation to assist our next step in the comprehension of everything and not just limited to an accidental local constant which is subject to natural changes. Additionally, current modern science is unable to make this next step as their physics has no mechanism available to them. The theory of the author set out in this book can change this rigidity of thinking and enable them to use a natural lawful mechanism of theoretical figures in all existing scientific equations, and henceforth make a **Flat world a Round one!**

Large Hadron Collider

The work being carried out in part at Cern the Large Hadron collider is spending vast sums of UK and European money in the hope to move a particle faster than the speed of light without realizing the physical rules why they cannot!

Should they consider this new theory will provide them with a totally fresh philosophical insight. And opportunity to widen their understanding of the results and modify their thinking regards the mysterious Higgs Field which does not exist. It is a natural function of the **dimension and mechanism of Spacetime (ut)!**

8 ABSOLUTE SPACETIME
IS A LICENCE FOR EVERYTHING TO MOVE IN
THE OTHER 3 SPACIAL DIMENSION'S

Newton said " *What is there in places where matter does not exist!* He was on the case – he knew something was missing, but simply could not imagine what it could be. Then Einstein completely discounted the idea of another dimension and since then, we have not really put our minds to it mainly due to Einstein's influence on us.

Let's stay of the subject of light a little bit longer rather than jump into a long list of other examples which may cause confusion. The illustration below shows how we understand light to move on an everyday basis. The image will not be new to you – but how I think it moves will be:

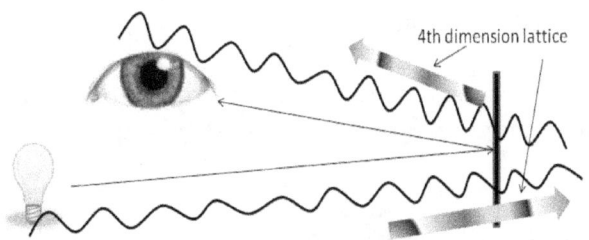

it is not the **energy** ability of photons in isolation to reach the wall so we can observe it: It is the ability of expansion, of Spacetime which is moving in all directions at the velocity of light .

We can see the black wall because the light which is being transmitted from the lamp travels through space at light speed, is reflected from the wall changes direction and continues its journey to our eye. We henceforth receive optical information that the wall exists. That is obvious and our eyes experience this behavior everyday without our minds (miracles which eyes are) even thinking about it. But if you recall in the earlier chapter it illustrated an example that this 4th spacial dimension may be comprised of an expanding lattice framework which is forever creating new space. It is the expansion of this framework which enabled the movement of light to reach the wall and hence so we can see it. Moreover, it is not just the fact that the framework is moving which enables the light to move, and determine the velocity of the light. The most important thing to consider is that it is this 4th dimensional framework expansion – is '**Creating** New **Space**' - at '**light speed**'. This enables the movement of the light and determines its velocity. The 3 other spacial dimensions are almost of secondary consideration. And in terms of dimensional importance this 4th spacial dimension should be considered the premier dimension because without it nothing would exist for the other 3 to occupy!

Extending our thinking about the meaning: '**Creation** of **New Space**' – **license for movement.**

The illustration on the next page gives us a better understanding of what this means and how it happens.

1. The man on the conveyor can only have his feet in space (meaning Spacetime) which is **being created,** in this case represented by the conveyor.

2. Also, in order that the man can move the conveyor belt must also be moving: **Spacetime must be expanding.**

3. If he runs so fast that he is travelling faster than the conveyor can produce New Spacetime – then his case is hopeless - and going nowhere fast. (**How can you occupy a dimension which does not exist)!**

4. How can you move – or indeed anything in space, where the Space has not yet been created for you to move in it!

The Mass of the man (**light**) is
being increased (getting infinitely heavier) as
he has no '**new space**' to move into. Despite the fact
we want him to run faster and we give him more energy

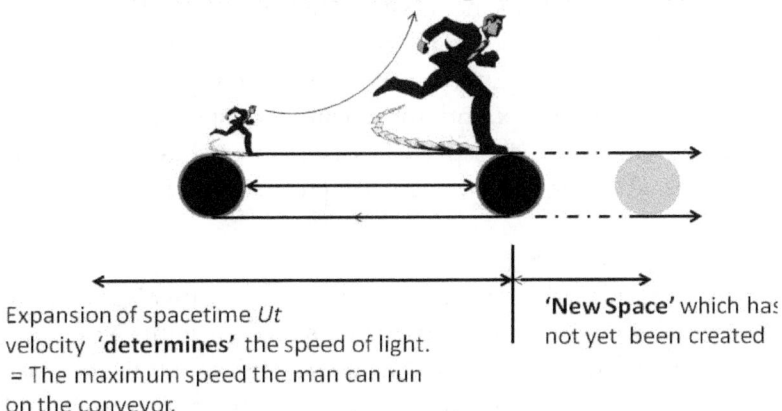

Expansion of spacetime *Ut*
velocity '**determines**' the speed of light.
= The maximum speed the man can run
on the conveyor.

'**New Space**' which has
not yet been created

If the conveyor (the universe expansion) stops expanding
all movement becomes impossible. The conveyor
Immediately dissapears and with it the 'End of Everything'

5. If the universe stops expanding we have to consider existing in Spacetime which is static. If such a thing happened how could we achieve the ability of any movement?

6. Think of this scenario not as the speed of light but 5 miles an hour which is easier to comprehend. Now the expansion speed of conveyor extension is 5mph. We are standing on the belt which is also extending at 5mph. As we walk along the belt each time we put a foot onto the belt – it has already moved away from us at 5mph of '**New Space**'. If the belt stops means no Space has been created and we can't move!

Think of this rule in 2 ways: **(a).** The conveyor belt has to keep moving and extending to create new space which we need to put each foot in front of the other. Should the belt stop then no new space is created for the next foot to occupy. (This is the best analogy I can think of - (you have to forget the idea that because the belt is moving you are going to be propelled along by it - this is not my point). The analogy is attempting to describe the fact that the conveyor belt movement is the manifestation of new space. You have a will to move and the fact that the belt is moving is providing new space for you to do that. If the belt stops no new space has been created and hence nowhere for the next foot to go. **(b).** If you have the will to run and accelerate along the conveyor belt you will attempt to run faster than the conveyors ability to extend and create new space. Whereupon you will have to wait until new space has been created - or waste a lot of energy in the attempt to move in to space which has not come into existence.

I hope that explains the meaning of 'license to move'. In physics it is well known that you cannot exceed the speed of light. You can pump zillions of Mega Giga Watts into a proton from an atom in a cyclotron the size of a planet, and forget about what the miniature HADRON collider can do. And the only thing you will achieve is a very expensive electricity bill and proton which weighs the size of a planet – but it won't go any faster! **You cannot move into a dimension which does not exist. You are obliged to wait for it to happen.**

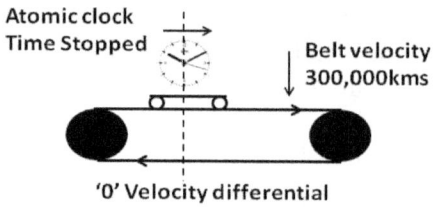

Atomic clock
Time Stopped   Belt velocity
                    300,000kms
'0' Velocity differential

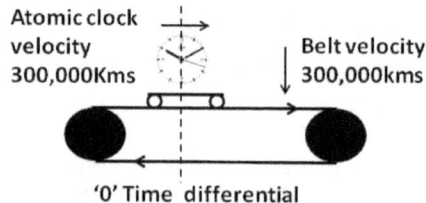

Atomic clock
velocity   Belt velocity
300,000Kms   300,000kms
'0' Time differential

Illustrated above the diagrams demonstrate the real existence of the dimension of spacetime. The top one shows the conveyor with a clock travelling with it. The speed of the belt is 300,000kms hour, should the clock move at the same speed as the belt it reads Time '0'. This in keeping with Einstein's prediction and is proven by experiments with atomic clocks. It is obliged to read zero (clock has stopped) because it is in velocity phase with belt.

In the lower sketch is shows the same scenario but this time we must consider the speed. When the clock read zero (clock stopped) it must be moving at 300,000kms.

The relationship is between the background velocity of the single dimension of spacetime and clock performance. They are both linked and is a coupling phenomena. In modern science since 1916 they realize that time dilates but have not been able to identify the cause. This cause seems obvious if you permit and recognize this coupling relationship between speed a time. Modern science simply will not recognize this because they think in a fixed 3 spacial dimensions + t proper time. If they were to finally accept that the 4th dimension exists then will solve all manner of current scientific problems. Such as the Horizon Problem, Comprehension of Time dilation and not have to keep looking for the so called God Particle. The God Particle is no more than the ability of the 4th dimension expansion which determines the ability of atoms to existence or not! Until they realize this science community will continue searching in vein. Which is a great pity because it is wasting huge sums of money and time which could be invested in real research to further understand our existence here in the universe.

This explanation satisfies my personal curiosity. Everything is determined by the expansion rate of the universe in our region of the universe and hence the absolute ability of things to move. And don't even try to move any faster than the rate at which absolute Spacetime is creating that dimension. This also affects the quantum and sub quantum world: The frequency of vibration available in atoms, or electromagnetic waves is also limited to the rate at which new space being created at 300,000km/s. If as stated earlier the absolute spacetime expansion 'ut,' varies up or down will in turn modify the physics of everything. In the new expression $E=m(ut)2$ 'ut' is the rate of local absolute spacetime expansion. So when cosmologists study distance regions in space-time the actual expansion in that zone may be at variance to ours. Namely, **E=m(ut)2** which may be either e=m<c² or e=m>c². It is proposed that science should find new ways of determining the value of E=m(ut)² in that area of investigation to fully understand and relate it to our framework of experience of 300,000km/s. I suspect however that this figure is uniform across our visible Hubble space zone. But we won't know for sure until we look for it – and it may vary – this may be one of its surprises!

9 END OF STARS

All stars in the night sky are nothing more than nuclear reactions. Due to a common centre of gravity of thinly distributed hydrogen atoms and other matter in space attracting each other they form huge spherical lumps of gas, dust and rocks. I and suppose eventually may include a NASA satellite or two floating around! When sufficient mass has been collected by this common gravity centre, the material is compressed under its own collective weight. Bearing in mind this mass of rocks, dust and hydrogen gas and NASA satellites may be 1,000,000 miles in diameter! Something as large as this of course will cause substantial internal compression. The result of this compression is that the hydrogen atoms it contains are fused (pushed together) to form a different bigger atom. The result of this causes the release of huge amount of energy (sunlight). Fortunately for us it is lovely and warm can give us a sun tan and ripen tomatoes.

The amount of energy released in this fusion reaction, is determined by Einstein's equation $E=mc^2$ mentioned earlier. And if you refer back to page 24 - a very substantial amount of energy is liberated from the tiniest piece of mass i.e. typically 1 gram in the example given will release energy to power an entire country for the best part of 1 hour. That equation reference applies the Mass x speed of light x the speed of light.

If the expansion of the universe creating new Spacetime **were to suddenly stop**, then the energy available in the above equation drops to zero. The net effect of that is that the fusion reactions in **all the suns** and stars which we can see, and can't see **would all stop instantly. The entire universe would simply be switched off.** Like a huge fairground attraction. It would not take 14 billion years to switch off – it would be instantaneous across the entire diameter of the universe – no matter how big it is and in the regions we can't see.

Why? You immediately ask – and a good question. The answer is simple. Referring back to our expanding chicken egg scenario earlier the **'Event Moment' of the Big Bang genesis** (not our proper clock time) has been preserved by the 4th dimension – Spacetime 'ut'. Which if you recall the event time is presented as symmetrical across the universe illustrated by the earlier simple sketch. So if the universe stops expanding for one moment – the entire fabric of the spacetime is immediately affected. The expansion is simultaneously ceased everywhere at exactly the same time. On the basis that there is zero time difference across the entire universe – it all exists in the same moment of time. No matter how small or how large it is. Think of this analogy - a rubber band stretched over a long distance (let's say 5 miles). Then cut it with a pair of scissors. The energy being conserved as you stretch it along its length is explosively released as soon as the band is cut. In this case the ends of the bands are held at all the outer faces of the universe. As soon as the band is cut immediately disturbs the entire length of the band. Even with a small length of 5 miles can you imagine the centre of the band at 2.5 miles? - shall you have to wait for the retraction to occur from the point of the incision? Not in my mind as the molecules of rubber are being uniformly stretched across the entire length uniformly. So as soon as you cut it and I am standing in the centre I will immediately know that you have cut it even if I can't see you cut it as you have disturbed the uniform tension recorded in the band.

One can think of Spacetime as a motive force for everything. And without it **nothing works** in the Euclidian/Newton/Einstein space - time of $x, y, z, + t$. The person who can stop the universe 'ut' from expanding – can end the universe.

The diagram on the next page illustrates its affect on a sun. The ability of Einstein's $e=mc^2$ to determine the energy emission from the suns nuclear fusion reactions. If the value of the absolute spacetime ut reduces it will immediately affect the reaction and availability of energy from that reaction. If it ceases all together then the sun will be turned off. On the other hand if the spacetime expansion increases will be just as bad for us as it would cause a substantial increase in the amount of energy cause and invite our extinction.

This gives us another way of looking and considering the physics of a sun – especially ours!

Expanding Spacetime *Ut* determines the speed of light and hence the ability of Einsteins E=mc2 equation. Hence The resulting release of energy which enables Suns to exist

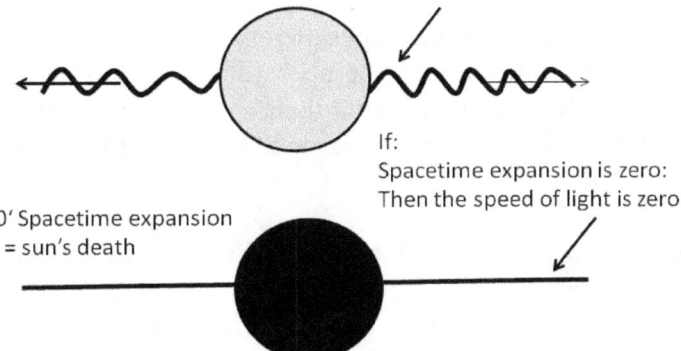

If:
Spacetime expansion is zero:
Then the speed of light is zero

'0' Spacetime expansion
= sun's death

10 TIME CRISIS

According to Einstein's Special Theory of Relativity, the effect of proper time (our wrist watch – or particularly atomic clocks – not spacetime) time is dilated (passes more slowly) as the velocity of that device is increased. This caused a well know paradox called the Twin paradox. It is only a mental experiment where the age of one twin will be at variance to the other if he goes for a fast trip near light speed. Then apparently when comparing their ages later the one who did not go for the flight looks much older or has died and his children already have grown up!

It is a well known and proven fact, that this phenomenon is completely correct – although we can't prove the twin paradox above, and for the following ideas is of no importance whatsoever. The important consideration is that proper time does actually pass slower according to Einstein's predictions. I am going to refer to these readings and devices as a means to demonstrate the real presence of the 4th spacial dimension being Spacetime, in the following way.

Atomic clocks are the most accurate proper time clocks ever produced by the hand of man. It is well known that you can have 2 clocks perfectly synchronized and put one on an aeroplane and sure enough it runs very marginally slower than its synchronized unit stationary (albeit the earth is moving) on the ground.

What the Special Theory states essentially is that as you approach the speed of light the passage of proper time will come to a complete standstill! This is the important point as far as considering the real affects **Absolute Relativity**. Atomic clocks measure microwave emission from atoms. These atomic materials are located in a suitable environment to ensure they don't get over excited and tick at precise intervals. Hence they record the regular ticks and passage of accurate proper time keeping.

If one of these devices is on board a velocity of light spaceship, and indeed the proper time recording drops to zero as they predict and expect. Then a few things are happening which we should be aware of regards our case for the presence and proof of the neglected 4th spacial dimension of Absolute Relativity.

1. The passing of atomic clock time has dropped to '0' zero because the atoms which it is monitoring which emit microwave (form of light) are **in phase** with the expansion of the local universe. Put another way: the emission of the microwave from the atoms which the clock is recording is travelling at the same velocity as the Spacetime expansion 'ut' in which it is moving through.

2. You could repeat this experiment in any part of the local universe and the result shall be the same. The clock would read repeatedly '0' zero time. Then if you recall the expanded chicken egg scenario the book states that the **'time in the entire universe is symmetrical'** – which means there is no proper time – albeit the proper clock time across the universe is '0' zero = the original **'Event time'** = **no time** – and spread over a huge distance – like our egg!

3. Our atomic clock has to travel at light speed in order to reach zero proper clock time. This is telling us the velocity of the movement of **spacetime expansion**. And the speed of light is of a secondary consideration.

4. We have several good arguments and case for the presence of the 4th dimension as Newton had dreamed of but could not fathom: **1.** the proper clocks will read **zero time** at light speed. **2.** It will read zero when used in any part of the universe when moving at 300,000kms. **3.** Atomic clock measurements and reliability is very high and creditworthy instruments. **4.** Various and numerous experiments have been conducted to prove Einstein's Special Theory of Relativity relating to proper time dilation using atomic clocks. **5.** The clocks must be at **light speed** to read zero. This is the same as case 1 above but we have to consider both. Time zero and Light speed. The expansion is an accidental 300,000Km/s and has a **constant Time symmetry of nothing! '0'.**

Moving on to another astonishing feature regards the effects of local spacetime expansion. You will understand the cases above and the concept that time is symmetrical throughout the universe as per the egg diagram and clocks all showing the same time in it.

Time in a bottle, your great grandfather is still alive as you read this!

If you can imagine that time is indeed symmetrical across the universe then some very unusual ideas are available. For instance, in the above cases we know that atomic clocks monitor proper time, and if they go fast enough time stops. This means they are monitoring both Spacetime proper clock time as zero and the Spacetime expansion is 300,000kms. They record proper time when travelling less than the speed of light (our conveyor). They monitor Spacetime which is zero when moving in phase with the expansion, **which just happens** to be 300,000kms. This is not a mystery now, but the state of science is still left wondering what this means. It appears lucid, and *'time'* for another illustration, as this is going to be difficult to explain – so let's start with a simple picture:

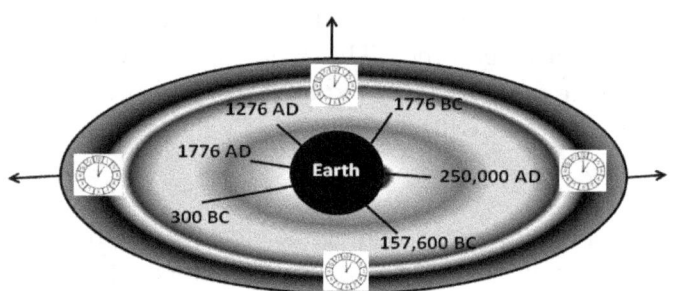

Spacetime expanding with fixed time
throughout the universe

So returning to our expanding egg, it shows the symmetrical fixed spacetime of '0' – the proper clock time all show 1 o'clock. Irrespective to where you are in the universe. In the centre – in fact somewhere in the universe just happens to be Earth. Around the Earth are illustrated random dates of time – some very early ones, some very future ones and a few in between. Spacetime has a symmetrical time frame which does not change and rests at zero. It is an **Absolute Relativity** to a proper time reference for all occurrences happening here on Earth. For example in the year 157,000 B.C. the time in the universe was '0'. Today is the 2nd of March 2012. The Absolute Relativity time frame is also '0'. In the future year 250,000 A.D. once again the spacetime will still be '0'. So what can we make of that? Here are some ideas: In all the time on Earth as measured in proper clock time – our experience of time, as far as the universe is concerned it is all happening at the same time always '0'.

Everything is happening against the same background of Absolute Relativity time frame! For example the life *time* of my own parents has happened, - as far as the universe is concerned, yes it has also happened – but it is happening during my life *time!* There is no difference in its time! Following me my son will have children, who will also have children - they will be born after my death. Again from the point of view of Spacetime it all happens at the same time.

More to follow: when my great grandchildren are born it will happen in my life *time* from the point of view of the background zero time framework. There is no change in time, everything that has happened and will happen all occurs at Spacetime zero. Obviously from our prospective of proper clock time passes and we have history. But when viewed from the perspective of Spacetime there is no history or future... Everything has the same Absolute Spacetime reference: time zero.

We regard our history, present and future by means of the proper clock time, movement of the Earth, seasons, when the shops open etc. This is a different reality of time and the one we only really think of, however, the absolute spacetime reference has never been introduced to us – it is another way of looking considering our lives! Our appreciation of time is given to us by what we see happens in the world around us, what we witness etc. But this is only 1 way of looking at the picture. The biggest part of nature is a complete mystery. We live in an x, y, z+t box and can't see outside it. This includes much of the science world. Otherwise they would have explanations for us.

From the point of view in '0' spacetime there is no change, it all happens in an environment of fixed time - the ultimate Absolute time reference. And to illustrate how small the entire universe is using the simple equation Speed = Distance / Time. To determine the speed at which the entire universe could collapse:

The observable Hubble Zone diameter of the universe:

8.08x10E26 meters in diameter
approximately 14billion light years.

1. If we use Sir Isaac Newton's simple equation:
Speed = Distance ÷ Time to find out how fast the
Universe will collapse

2. Speed = 8.08x10E26 meters ÷ 0 seconds
= 8.08x10E26 meters in 0 seconds

3. The above equation means that **the entire**
Universe will disappear in less than 1 second **0 seconds!**

The thought of all the suns becoming extinct in the same second is
alarming enough. But this theoretical event shrinks even that vision!
To illustrate the meaning above with a picture, please note the
following image to help with this understanding:

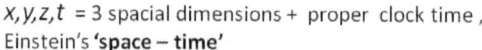

x, y, z, t = 3 spacial dimensions + proper clock time,
Einstein's **'space – time'**

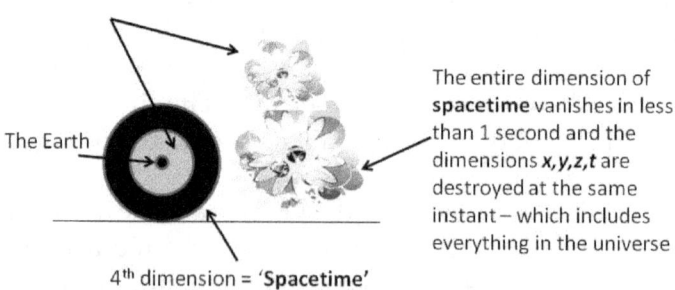

The Earth

The entire dimension of
spacetime vanishes in less
than 1 second and the
dimensions x, y, z, t are
destroyed at the same
instant – which includes
everything in the universe

4th dimension = **'Spacetime'**

I am not a harbinger of doom! If we think about it logically, we
either have a dimension or we don't! We can't have a part of
dimension – it either exists or it does not! And if we are going to lose a
Euclidian dimension such a width I can't imagine a scenario where

gradually the ability and presence of that dimension vanishes slowly! That would be completely preposterous. It either exists or it does not. So when thinking about the ability of the 4th dimension of Spacetime likewise it exists or it doesn't!

So if Spacetime vanishes in less than 1 second - an incredible thing to imagine I know, then we just have to accept this as a possible reality. **The universe is lot closer than we think!**

Conservation of Proper time

Since the beginnings the entire universe spacetime 'ut' has been expanding. Passing through us and every single atom of this planet and all other cosmological bodies, suns galaxies black holes etc. Then should the fabric of spacetime be composed of a superstring lattice means that all information and proper time has been recorded. This can be explained by the following sketches.

All knowledge is recorded in spacetime

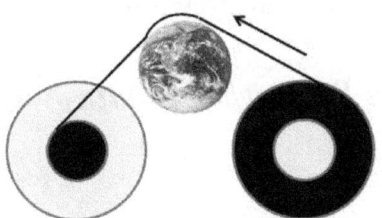

The fabric of spacetime expanding and passing through the Earth and recording our proper time experience of life. Every atom every flower - Everything

New space is being created and passing through us at the current rate of 300,000kms second for second for our entire lives. All our activities, conversations, hand and eye movement is recorded faithfully just as if it were video recording with voice. In true 3D colour and the length of the tape would be in our life time 70 year's x 300,000kms. A very long tape should you wish to calculate that! That would be the amount of New Space 'ut' created in your lifetime. But of course I hope you will live a lot longer than 70 years.

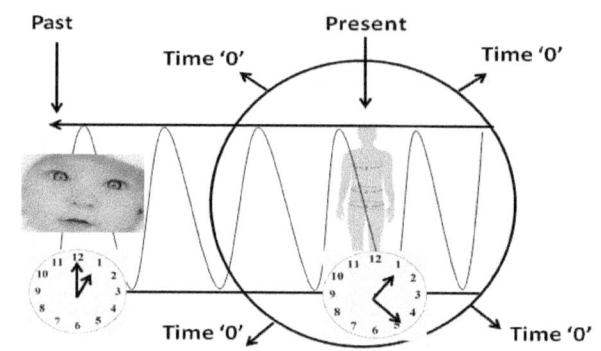

Expansion of spacetime in the experience of one persons life: 70 years x 300,000kms. Recording your history and conserving the passing of proper time. With a spacetime 'ut' background time of zero.

All information is recorded since the very beginnings of the creation of the universe in the medium of ever expanding spacetime.

Sir Isaac Newton

" What is there in Space where matter does not exist ?

Edward Johnson

"Absolute Relativity"
Matter exists because of it, and determines its character

11 DEVILS ADVOCATE

Scale of light speed

The speed of light as we experience it is 300,000kms per hour. At this speed it takes light to travel from the Sun to Neptune in 4hours. In this time we can cook Sunday lunch eat it and watch a movie. When we think of the speed of light in the human scale it appears impressively fast. However, the size of our sun, and that of our entire solar system is sub microscopic in terms of scale of the universe. So I think of it as being **incredibly slow**.

Furthermore, if we depend upon the known speed of light to rely on the velocity of expansion of the entire universe it just does not fit, - the 'scale of ratios' are troubling. By that I mean, in order for the current expansion to continue it will presumably have to have a velocity substantially greater than the known speed of light at the periphery. I could easily imagine that the periphery – that is the zone which is creating new outward space for our inner space - must be moving in the region of x1000++ times the known speed of light. Even this would not be a substantial number when thinking of the scale and the huge volume of the entire universe. Furthermore, if the speed of the expansion is only a few times that of the speed of light, I think the universe would have collapsed millions of years ago being so slow.

To drive the current and historic expansion is using unbelievable quantities of energy, and whilst that energy persists, and sufficient to continue to enlarge the universe – then Spacetime 'ut', will continue to provide us a dimension to exist in. When the energy at the periphery is insufficient to drive this expansion – bearing in mind the larger the volume becomes the more energy it requires, then a few of the following events may occur. It will not be providing new space which means it will affect the speed of light in our zone and all across the universe diameter. The result of that has been stated earlier – all the suns may expire immediately. Then should Spacetime remain intact life will be extinct in any event as without the expansion nothing can move. So that outcome is just as bleak as if it disappeared in less than one second! Then, whether or not the universe collapses due to its

45

common mass centre or effects of dark matter etc which now seems most unlikely, is of little importance as the damage as far as life is concerned is done. On the other hand if, and when the peripheral expansion is unable to continue expanding it may well cause **instantaneous total destruction** described earlier. One may think of this scenario as: the universe is a balloon as some people think. That is in order to blow up a balloon you have to use lung energy and the will to blow it up. If you are short of energy and the will to do that then the balloon will contract very quickly in front of our eyes. Then again, if you over extend the fabric of the balloon where it cannot withstand the pressure of you blowing in more air – it makes a nasty Big Bang! The result of that is the balloon vanishes quicker than we can observe it? It has lost the inner dimension to support the periphery!

Trouble with Strings

In the earlier chapter I have proposed that the concept of Spacetime is constructed from Superstrings. Of course I have no idea whether it is construction from these materials or not. However, these materials are a possible candidate and String theorists are short of a dimension to make any kind of sense from their postulations. So this may be the dimension to solve their problem. Additionally, the **normal** and easy **3** spacial dimensions exist which, we take for granted so why don't we add this further **one** which could provide an answer to many things in physics and biology. Albeit lumps of it may have to be re written!

Horizon Problem.

Cosmologists have one HUGE problem with their appreciation with of the universe, which, is known as the Horizon Problem. Put simply when they measure the background temperature of the universe it appears that it is homogenous. Essentially what they are saying is that the temperature is the same everywhere! Then if you consider the huge diameter some 40 odd billion light years – one would expect regions of thermal non uniformity (*different space temperatures everywhere*). That would seem logical on the basis: how can the temperatures so far apart communicate thermal differential information? "I am cold can you give me some heat" Across a staggering 40 billion light years. There is no apparent medium for the heat to balance out across that distance. And even if there was heat transmission it would consume to much **proper time** to move across that enormous void. And in taking so long to transfer that heat to create the equilibrium –Then in the meantime the other side of the universe would shiver to death!

An example: my kitchen is cold but my living room is warm – that is easy to understand because my kitchen is contiguous with my living room (*next door*). And I am putting heat into my living room and none into my kitchen which explains the thermal difference.

That's easy to understand, but my kitchen in the universe is 40 billion light years from my living room and temperature in both rooms is the same! **That is the cosmological problem**! Logic compels us to declare there is no way they could be the same – that is ridiculous as they are so far apart. **Well they are the same** and we don't know why! *But wait one minute....!*

How can the thermal conditions be the same when they are so distant from each other – that sounds crazy- there must be an explanation?.

Well there is: Since Einstein had discounted the concept of an Absolute Spacetime, causing Newton to turn in his grave, we have been unable to answer these questions. Now once again we can open the door of comprehension that spacetime 'ut' exists as a single dimension of zero time existing across it, **the distance becomes a complete irrelevance.** The thermal information in this case is transferred instantly across a plane with a zero time spanning it – and the distance becomes invisible as if it does not exist – which it doesn't in the plane of a single dimension of time '0'.

A challenge for Atomic Physicists

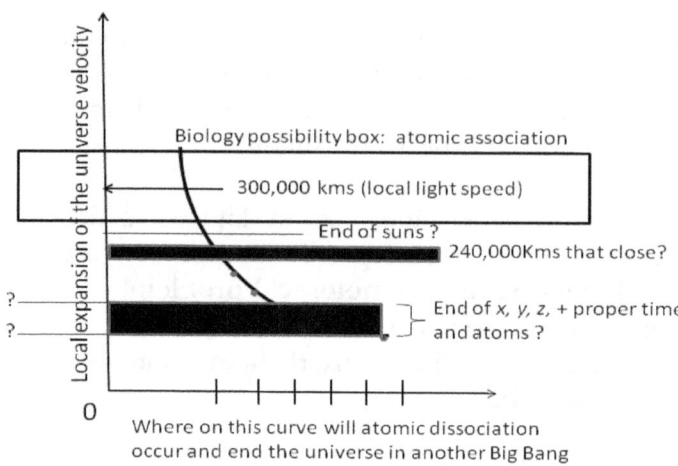

Where on this curve will atomic dissociation occur and end the universe in another Big Bang

The above sketch shows the left vertical axis speed of light from zero velocity – to speeds greater than the known speed of light by the direction of the upper arrow. In what I have named the biology box we know the speed of light to be 300,000km/s. We enjoy the existence of atoms, the taste of chocolate which, is made from atoms and the basis for all known physics. Physicists live in a world of the stated 3 spacial dimensions + proper time.

So here is the question: If they theoretically adjust their calculations to permit for a **variable speed of light** (expansion of the universe 'ut'), at what point will the atoms dissociate? They will be able to determine the answer and provide us with a theoretical lower value for the speed of light at which this occurs. In doing so they would provide us: the rate of expansion necessary to initiate the 'New' Big Bang'.

This answer may be zero or some velocity between zero and of course the Biology Box. If their answer is 285,000Km/s for example then that would give us a new outlook of the universe, on the basis that we are right on the fringe of biological existence. I.e. should the expansion of the universe be reduced by a relatively small amount will cause the end of everything! We may not know it but life is only just possible? And forget about the distance the Earth is from the Sun – as we have a new parameter to consider! Then should the conditions change within the absolute spacetime dimension we all vanish back to energy instantly – and no chance to say goodbye! All proper time and history is vapourised!

Atomic physicists may reject this notion on the basis that atoms have what is known as the lowest ground state. Determined by the ability of absolute zero temperature, however, on the basis that liquid helium is quite happy to be a **perfect liquid** at minus -273.15centigrade (absolute zero temperature)! This means that whereas they have determined a number for the lowest possible temperature, it does not account for the fact that the atoms are still busy vibrating.

Liquid helium might be very cold but from the point of view of its atoms they are **not** frozen solid their energy dynamics can be taken a lot lower even at this temperature. Not by reducing the temperature which we know is probably not possible. But rather take the view that their lowest energy state is determined by the expansion velocity of the universe passing through it.

To explain: as previously stated the sub atomic components constructing atoms are only able to vibrate in the medium of spacetime moving at 300,000km/s.

If however we state that the expansion has a lower velocity – then they will be obliged to vibrate at a lower frequency.

If this ground state continues downward, there must be a point where the ability of an atom to remain an atom is not possible. Then under those conditions they will dissociate and with it release energy using $e=mc^2$ (**actually E=mut²** on the basis that Einstein states '**C**' as a God given constant. I think that is a luxurious claim). And the calculation of the released energy is determined by the lower rate of expansion – not 300,000km/s.

The scary bit is the number which atomic physicists will present us with. For example the current speed of light is 300,000km/s. which, enables the Biology Box to exist and established known physics. If they state 289,999km/s, = time to prepare for Absolution! But we should not have to worry about being made deaf or blind by the Titanic explosion – our bodies comprise atoms. A typical atom bomb uses approximately 2.5kgs of plutonium and is equivalent to 50,000 Tons of TNT being detonated. The average weight of a man in the UK is 84Kgs. Which means our individual bodies will release approximately 84 x 50,000 Tons = 1,700,000 ton explosion. All our individual bodies have the ability to become massive atomic bombs if the velocity of spacetime decreases.

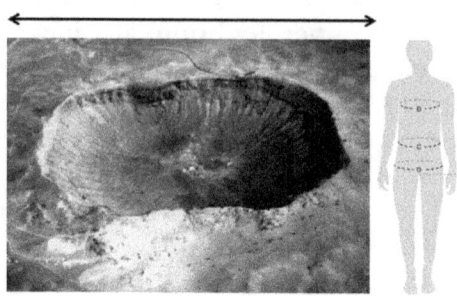

We are potential atomic bombs.
All the mass in one person will liberate
energy equivalent to a 1.7 Megaton atom
bomb sufficient to cause a crater
1 mile in diameter.

This 1 mile crater above represents the amount of energy released from a 1.7Megaton bomb. This image is actually the result of a meteorite impact some 50,000 years ago. This is the amount of damage all the atoms in 1 person's body could achieve upon release of that energy should the rate of universe expansion slow down and cause the dissociation of our personal atoms in our body.

12 NEW UNIVERSES

As stated earlier the plausible actual velocity of the universe periphery should be moving at speeds substantially greater than light moves in our part of the universe. At these huge theoretical levels of momentum (movement) ordinary quantum materials (particles & atoms of mass) simply could not exist. The periphery being nothing more than an energy field stretching the 4th dimension into a void where space does not historically exist – although we can't be sure of that – and our universe may well be invading another!

As the periphery moves out into the void and creating Spacetime behind it, it will experience a lower momentum. Resulting in a lower momentum and energy will fall off significantly. This will cause energy in that zone to coalesce into what is known as baryonic material (standard model particles and hence atoms). As soon as that occurs we then have the building material for new galaxies, suns and planets.

Please note the diagram below showing the possible radial distribution of energy across the entire universe. As the ultra high energy field of the periphery moves out creating new space. The zone behind it has a lower momentum and coalesce new matter.

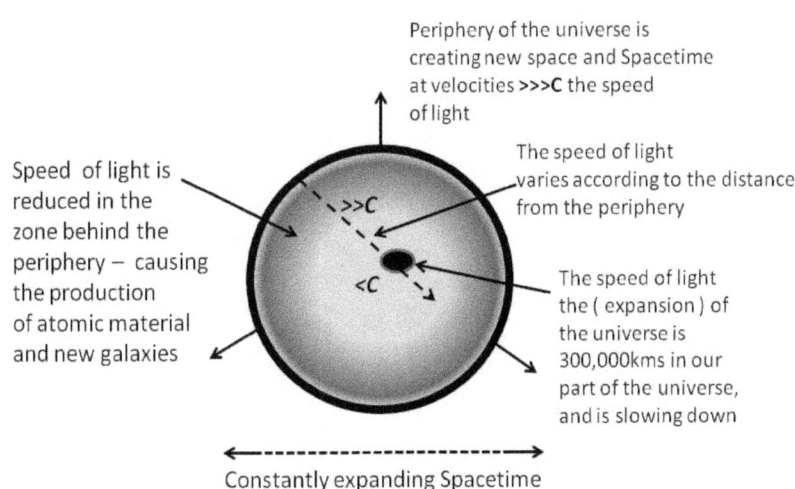

Periphery of the universe is creating new space and Spacetime at velocities >>>C the speed of light

The speed of light varies according to the distance from the periphery

Speed of light is reduced in the zone behind the periphery – causing the production of atomic material and new galaxies

The speed of light the (expansion) of the universe is 300,000kms in our part of the universe, and is slowing down

Constantly expanding Spacetime

Considering the above effects of expansion, one is given to understand that a very substantial quantity of universes similar to our Hubble zone may well be produced constantly. And inhabitants of those universes may think of themselves as being created by a Big Bang? On the basis that the periphery which created them would long since have disappeared by the time evolution has permitted intelligent life forms the ownership of a telescope.

It is also proposed that the background Cosmic radiation with particular reference to Ultra High Energy particles are the left over materials as a result of the continuous conversion of the energy to mass in the baryonic zone behind the peripheral expansion creating those new universes.

13 HIGGS BOSON GOD PARTICLE AND THE HADRON REACTOR

The publicity and popularity of the HADRON Collider has global interest. It is a very interesting project and the science being investigated. Most of us don't really understand it other than realising it involves smashing on bit of an atom into another at speeds close to the locally available speed of light. And apparently they seek what is known as the Higgs Field. This apparently causes mass less particles which move through it to obtain mass. I am not sure how that fits with the concept of this 4th dimension except in the dynamics of a black hole. However, this book presents the notion that nothing can penetrate a space which does not exist. So if a Higgs Field does exist it is inferred that it must be moving at, and or being created at, the rate of spacetime expansion. You cannot fill a dimension until is available.

Mass of Bodies & Gravity

Moreover, the **Rest mass** (*mass which is theoretically **not** moving! - it collects more mass the faster it moves or gets hot. Physicists tend to ignore this as it invokes relativistic mass consideration and causes huge problems with x, y, z, + t space-time with a frame of reference which is invariably moving here on Earth*) of a body has a numerical value determined by the rate at which this dimension is moving through it. Should the rate of expansion vary this will immediately produce new rest mass value.

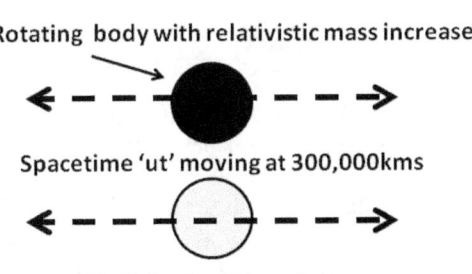

Rotating body with relativistic mass increase

Spacetime 'ut' moving at 300,000kms

Static body with rest mass

The measured gravity of any cosmological body, planet, sun or galaxy etc is a direct cause of the rate of 'ut' expansion through it. The subsequent secondary effects are the measured space-time field curvatures existing around them.

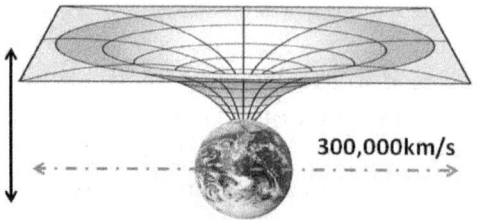

**Space-time curvature caused by the
Earths gravity in an expanding 'ut' field**

In the above diagrams it illustrates the net effect of (**space-time x,y,z + t**) curvature caused by the strength of Earth's gravity when '**ut**' is expanding at the current rate of 300,000Km/s.

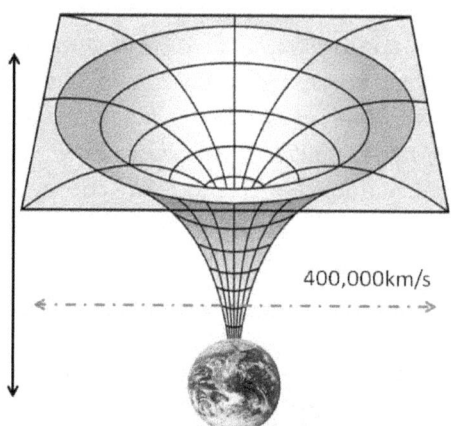

Space-time curvature caused by the
Earths gravity in an expanding **Ut** field

In the second lower diagram it illustrates that should the local rate of 'ut' expansion be a higher value for instance 400,000Km/s, the Earth will of course will be marginally smaller but its affects on the space-time curvature will be significantly greater. This demonstrates that the value of gravity around any cosmological entity or body is subject to the speed at which the expansion of the 4th dimension ut is moving through it. The relativistic mass and hence gravity is a direct function of its rotation in the plane of spacetime expansion. Should the 'ut' value change upwards the Earth will become heavier and hence reduce its diameter, and of course rotate faster. This is the same rule for black holes at the beginning of this book.

Albert Einstein

"Nature shows us only the tail of the lion. But I do not doubt that the lion belongs to it even though he cannot at once reveal himself because of his enormous size."

Edward Johnson

The lion is the numerical value 'ut'.
The tail is 'C'.

14 EVOLUTION
OF SPACETIME

1. Sir Isaac Newton, was aware of spacetime and unfortunately could not provide us with insight into its applicability in science, and referred to it as the Aether in the universe.

2. Victorian scientists continued to look for the Aether without any conclusive results. They assumed it was a luminiferous wind which was the medium for the conduction of light in a vacuum.

3. Einstein completely discounted the applicability of Spacetime prior to his 1905-1916 work and focused on the known dimension's of Euclidian space + Al-Jayyani curved space and of course Newton's historic work on gravity. And invented an ingenious set of scientific laws. But unfortunately these only refer to laws in Euclidian space + gravity. Confusion occurred, and continues between the idea of Spacetime and space-time mainly due to Einstein's exclusion of any importance of Spacetime. He knew it existed but could not provide us any insight. And he made a decision - it did not participate in his vision of the universe – space and time. – Leaving Spacetime misunderstood.

4. Notwithstanding, there have been scientists who have tried to establish that Einstein space-time is time symmetrical. Unfortunately they were aiming at the wrong target. Their work once again is based upon the behavior of atomic and sub atomic particles. Which means their experimental work was confined and focused on Einstein and Euclidean space and discounted the notions of real Spacetime effects.

5. Now we have some new ideas centered on String Theory and Higgs Boson fields. This new science group are seeking the answer to the same question as Newton – what is space made of and how do we include it into our collective perspective of the universe.

I trust the ideas presented in this book provide the insight which we seek. And the understanding that the 4ᵗʰ Spacial dimension does exist, and further develop our understanding of the universe and the world.

Absolute Relativity
The Theory of Everything:

1. (Ut, x, y, z): The 4 spacial dimension's of the universe.

2. E=m(ut)² : The energy mass ratio where 'ut' is the velocity expansion of the universe.

3. Ut: = The speed of the universe expanding in all directions.
The Primary spacial dimension

15 THE AUTHOR

Edward Johnson was born on the 16th April 1952 in Gillingham Kent United Kingdom. His father was a Royal Naval Officer and had an appointment to be stationed in Singapore in 1958 to run the Naval Hospital. The author attended the local Naval School in Sembawang on the north shore of the Island facing Malaya. Not a great deal of academic work was conducted in this period, and most of this time was spend in the Naval base swimming pool or wondering around the Chinese and Asian Kampong's, in search of leaches, insects and snakes! In 1963 returning to England was reintroduced to the concept of education, rain, wind and snow, attended a cold granite primary school in South Brent Devon, and forced to wear a cap and short grey trousers.

Completing his secondary education commenced a course of Horticulture at Dartington Hall School also in Devon. He wanted to follow his father's influence, with ambition to become a Royal Navy pilot. Before joining the service he worked as a builder's labourer to earn enough money to visit the Exeter flying club to earn his private pilot license in preparation and hope to become a professional pilot. After 6 hours of instruction was sent solo in Cessna 150 aircraft with the instruction to fly one circuit. Instead he took this opportunity to take an unapproved flight around the local area. Causing substantial concern on the ground, they thought he was too scared to land! Correct!

He attended the RAF Biggin Hill pilot aptitude test on 3 occasions. On the final occasion he passed, but then have his application turned down by the Admiralty Interview board. He joined the Royal Navy in any event and spent his career on numerous, mechanical, electrical, aviation courses, including a 2 year course involving the Russian Language during the Cold War.

In 1978 during his service with the Royal Navy was interviewed by the Army to work in a Northern Ireland video shop during the IRA problem and ordered to grow his hair long. That was ironic he thought as he just had it cut to join the service. In April 1982 joined HMS Invincible 820Naval Air Squadron to participate in the Falklands War.

During this period he spent 6 months at sea along with his shipmates including HRH Prince Andrew, interrupted by visits to the Island, to witness what a mess it was in!

In 1988 he was awarded £100,000 R&D grant by HM Government whilst still serving in the Royal Navy, on the basis that he and his science partner had a new concept to deposit transparent conductive thin films onto glass for the production of flat panel televisions – which at that time did not exist. Since that time has invented various industrial devices including a novel means to separate particles from gas.

He is semi retired, and now lives part of his time in Burgundy France, where he is studying the range and species of abundant wild orchids around the villages. He is a keen artist, inventor and musician, and has a love of nature and the countryside. He has travelled the world and is a member of The Royal Institution of Great Britain. He has a son called Thomas Edward, being his only child aged 17 years.

Thank you for thinking about these ideas and purchasing my theories. It is not so gloomy Spacetime records information about *'everything'* - you, me, our families, past present and the future like a huge natural Hard Disc Drive second for second since its beginning.

We can be excited by the possibilities of what that means!